MW00769264

Antarctic Expedition

First Collection of Algae by SCUBA

Richard Fralick, Ph.D.

THIS BOOK IS DEDICATED TO KATY, MY WIFE AND BEST
FRIEND, WHO HAS PUT UP WITH ME ALL THESE YEARS

AND TO:
THOMAS AND EMILY,
HANS,
OLIVER AND FINN

Table of Contents

Introduction 7

Chapter 1: The Beginning 11

Chapter 2: The Journey 29

Chapter 3: The Melchior Base 50

Chapter 4: The Algae 77

Chapter 5: The Animals 86

Chapter 6: Dangerous Situations 108

Chapter 7: Other Bases 122

Chapter 8: Diversions 132

Chapter 9: Leaving Antarctica 139

Chapter 10: Epilogue 150

Appendix A: Antarctic Treaty (1959) 161

Appendix B: Maps 164

Appendix C: Glossary of Spanish Terms 166

Appendix D: Bibliography 167

Introduction

Why did I write this book? After 40 years in academia, including teaching and research on marine algae around the world, I decided to write a book describing my work for my grandchildren and former students. My goal is to inspire young students to protect the marine environment.

The first requirement is to learn about the ocean and the organisms that live within it. The magnitude of the ocean is vast and needs to be studied intensively in the future. Education has become very expensive and technology has become more complicated and precise. But the challenges are real and still remain exciting. The route to facing challenges in education boil down to defining your passion and developing a skillset to develop that passion. For me, I was rewarded by using my SCUBA diving skillset and adopting an interest in marine algae.

It took me a long time to obtain the education I needed to develop my knowledge base in marine biology. Sometimes I was only able to take one course per semester. I took every opportunity available to learn what was necessary about marine algae. I was also lucky enough to enjoy the support of my teachers, professors and fellow students.

This book exemplifies how I was able to participate in the early stages of underwater exploration for algae. My fascination with marine research took me on a journey at a young age to Antarctica. Antarctica was an eye opener for me and the opportunity to be a student of algae under the guidance of world-class researchers from Harvard University was an opportunity beyond belief.

The Harvard professors have passed away and although numerous research papers were published regarding the algae we discovered, no accounts of the day-to-day activities of our time on the Antarctic Peninsula were published.

Each of the two Harvard professors, Lamb and Zimmermann, had given me copies of their daily journals for the entire length of the expedition. I combined their journals with mine and was able to put together some stories about our work in Antarctica via three different perspectives.

I also pointed out how our expedition reflected the best parts of the Antarctic Treaty of 1959. I truly hope you enjoy the stories and become inspired to follow your passions by developing your skillsets. There are opportunities to become involved in the USARP—The United States Antarctic Research Program. For more information, you can contact the National Science Foundation and possibly visit a base as a worker or intern.

Chapter 1
The Beginning

"I had a dream when I was 22 that someday I would go to the region of ice and snow and go on and on till I came to one of the poles of the earth."
Ernest Shackleton

THE RIPE OLD AGE OF 79 is a great time to summarize the unusual pathways of my life that somehow got me to this point. It all started on July 27, 1937 when I was born at Boston Floating Hospital. (I wonder if the floating part had something to do with my life on the ocean.) I spent most of my childhood in Somerville and in St. Clement Elementary School. By the time I was ten, my father had left home and so my mother became a librarian.

I started Saint Clement High School in 1952. I was a pretty good student who did well in my studies especially in math, science and Latin. I had a good memory and could pass exams without much study time. I also liked to write book reports and term papers. By my sophomore year I played football, basketball and baseball. I wasn't destined for any future or scholarships in sports, but I did learn the rules of the game.

During my senior year of high school, I worked for New England Telephone & Telegraph as a cable splicer and installer repairman. I made a routine repair call to the house of a woman named Ruth Elder, a theater professor at Tufts University. After hearing my voice and diction during a completions "call in" to the central office, she suggested that I consider going to college. I met with her at Tufts and was introduced to the admissions officer, who, upon Ruth's recommendation, urged me to fill out the application and submit it with my SAT scores as quickly as possible. I then took affordable courses at Boston University and Suffolk University in Boston.

One rainy weekend, my close friend Kaz and I decided to skip our ski weekend and headed to Boston to see the underwater movie, "The Silent World" by Jacques Cousteau. This is the first underwater movie using untethered SCUBA (Self Contained Underwater Breathing Apparatus) divers and underwater cinema. It really made a profound difference for both of us. The next day we went to Brines Sporting Goods store in Boston, purchased mask, snorkel, and fins, and got the number of George

Dainis, a SCUBA instructor from Somerville who taught at the Cambridge YMCA.

Kaz and I took the certification course and began diving at Halibut Point in Rockport, MA on weekends in the spring. Kaz and I were committed to SCUBA diving for the rest of our lives.

Diving was ecstatic! Visions of treasure still danced in our heads. We learned about decompression sickness and Nitrogen narcosis and Boyle's Law. We also learned that the weight of the atmosphere (atm) had a pressure of 14.7 psi surrounding our bodies. Psi stands for pounds per square inch. This is a value applied to the pressure (or weight) of the atmosphere around us. The value of 14.7 psi is the standard weight of the atmosphere at sea level. As we dove deeper in the ocean we learned that the 14.7 psi, or 1 atmosphere of pressure, increased every 33 feet of depth. Thus, the surface is 1 atm, while 33 feet deep is 2 atm, and 66 feet deep is 3 atm or 44 psi of pressure, etc. There were limitations in the length of time at that depth that we could work on the ocean floor. A typical dive was about 30 feet at 2 atm pressure, for 40 to 50 minutes.

We dove every weekend and on some weekdays during the summer. In the fall, George Dainis asked if we would help with his weekly diving class at the Cambridge Y. This was great! What an opportunity. We were diving instructors, which was very cool. Often, we would bring students along on our weekend dives at Halibut Point in Rockport, MA.

In this position of diving instructor, we met many interesting people, including Lloyd Breslau who was a graduate student in 'Doc' Edgerton's strobe lab at MIT, and Dr. Mackenzie Lamb, or Mack. Mack was the Director of the Farlow Herbarium of Cryptogamic Botany at Harvard University. His research was primarily on Lichens, an organism composed of an alga and a fungus living in a mutually beneficial status. The alga typically undergoes photosynthesis and produces glucose.

We brought Mack to Halibut Point in Rockport to dive with us on weekends. As it turned out, he became pretty excited at being the first scientist to actually observe the New England marine algae *in situ* at depth. In fact, he was one of the first biologists to study marine plants *in situ* in the Western North Atlantic.

Mack was so interested in the algae and proved to be such a skilled diver, that he offered me a job working for the Farlow Herbarium as a Technical Research Assistant. I said YES!

My major working task was to go diving with Mack on weekends and collect algae at different depths, both for the herbarium collections at the Farlow and for a course on algae for Harvard undergraduates. My role as one of Mack's diving instructors led to this job at Harvard. In fact,

I was actually getting paid for something I really enjoyed doing as a hobby. This was an outstanding opportunity for me to develop an interest in seaweeds and their ecology at a time when marine ecology was just getting started as a serious scientific discipline.

When I first started working at the Farlow Lab, I learned how to identify, process and preserve marine algae, in addition to my weekend algae collecting duties. As time passed I participated in a variety of expeditions to places like Bermuda, Canada, and Venezuela.

During each trip, I honed my skills farther, learning more about SCUBA and algae. Dr Lamb and I got funded for a student exchange program with Universidad Central in Caracas. This involved bringing students from Cambridge to Caracas and then the Venezuelan students back to Harvard to dive and study algae in Rockport and Cape Cod Massachusetts. The results of this exchange program were spectacular and I gained a wealth of experience.

The next two years were spent diving, helping in the lab, taking courses and preparing for the next big expedition to Antarctica. This required tons of reading, organization, and evaluation of diving and survival equipment, and a year of testing specialized equipment, which included a great deal of winter diving in Rockport and Cape Cod. In Antarctica, we would be left on a small island with a small support staff, completely on our own for many months, so we had to prepare. I can't begin to explain how excited I was to be heading to Antarctica on a SCUBA expedition to collect seaweeds.

The National Science Foundation funded the Antarctic Expedition, officially called *Bio - Harvard - Melchior*. Dr Lamb, aka Mack, chief scientist and director of the Farlow Herbarium at Harvard, was in charge of the expedition, and I was in charge of all the field operations including equipment procurement, maintenance, shipping, and setting up the dive teams.

History of Modern SCUBA Diving

Self-Contained Underwater Breathing Apparatus, or SCUBA as we know it, requires a tank of compressed air with a regulator to reduce the air pressure from over 2000 psi to around 100 psi so it can be used in breathing underwater. Experimental types of diving apparatus were beginning to be seen in the late 1800s. By 1891, the US Navy Dive Manual discussed the development and utilization of the MK V Deep Sea Diving Dress. This was the standard rounded brass helmet, canvas/rubber suit, and heavy lead weights suspended at waist height and also made into heavy weighted shoes. It wasn't until 1943 that Cousteau and Gagnan patented the demand regulator which regulates compressed air 'on demand'.

Although humans have been using goggles and variations of a snorkel for well over a hundred years, it wasn't until 1943 that SCUBA was invented and commercialized. Prior to the SCUBA invention, various hard-hat divers in pressurized suits enabled underwater ship repairs and scientific observations. SCUBA meant that divers with an air tank on their back could easily work underwater untethered.

Since SCUBA was invented in France it evolved locally for several years before it crossed the Atlantic to coastal America. With SCUBA available in a few East and West coastal areas, protective clothing was required to cope with SCUBA diving in cold ocean waters. A physicist from Scripps Institution of Oceanography in California is credited with the invention and development of the neoprene wet suit. The early wet suits were made by cutting sheets of 3/16-inch neoprene according to a pattern, which could be contoured (with a few measurements) to be a good fit. The pieces were joined together and glued with some black rubber cement called 'Black Magic'. Most divers made their own pullover shirts, pull-on pants, head covering, socks, and gloves, all out of neoprene. This outfit kept us warm in cold (65° max) summer ocean water.

The mission of our project was to make a detailed study of the marine vegetation (attached marine algae) and other plants such as lichens, mosses, and grasses occurring along the coast of the Antarctic Peninsula and its adjacent islands. This study comprises the taxonomy (classification) and ecology (habitat relationships) of the organisms concerned. Our working team consisted of five members, four of who were certified SCUBA divers. This international project was collaborative within the guidelines of the Antarctic Treaty which provides unique opportunities for research in Antarctica.

A Botanical Survey in West Antarctica

Many people who have visited the Antarctic Peninsula were whalers. We were among the first to visit for botanical purposes. Papenfuss was an algae researcher interested in Antarctic algae, but 50 years before us. He did not have the technology to thoroughly study the algae and other plants the way that we did.

PRINCIPAL INVESTIGATOR: I. MACKENZIE LAMB, PHD

Papenfuss (1961) showed that early research on the marine algae of Antarctica indicated rich benthic vegetation. An ecological study of the Antarctic marine algae was initiated by Scottsberg (1941) during the (pre-SCUBA) Swedish South Pole Expedition and continued by Lamb in 1944-1945. Several researchers including Dr. René Delépine and Dr. Mike Neushul have made subsequent contributions to our knowledge of the ecology of Antarctic algae. In spite of these important contributions to our knowledge of their taxonomy and ecology, much work in both disciplines remained to be done to present a comprehensive and reliable account of the

Antarctic marine algal flora and its aspects—taxonomy, life histories, geographical distribution, and ecology.

The object of our field work was to collect and study marine and terrestrial plant life of the region, paying particular attention to the attached marine algae of the coastal slopes by using SCUBA diving techniques to collect the plants and make ecological observations *in situ.* It was also planned to make an immediate study of the living material thus obtained in the laboratory and take color photographs (including photomicrographs) to illustrate morphological and anatomical features for final publication. Whenever possible, cultures will be made to throw light on imperfectly known life cycle phenomena. The use of SCUBA by an organized research team was unknown in Antarctica thus far, though Neushul and a few others made spot dives in some Antarctic waters.

The Botanical Survey of the West Antarctic Peninsula was completed successfully and safely with no injuries to the divers. Thirty-one genera and thirty-four species of benthic marine algae occurring on the coast of the Antarctic Peninsula and adjacent islands were described and illustrated. The algae were collected by the use of self-contained diving techniques (SCUBA) during the austral summer season of 1964-1965, the base of operations being the Melchior Islands off the west coast of the Antarctic Peninsula (64°19′S and 62°57′W) where all facilities for the expedition were made available by the Argentine Hydrographic Service.

The NSF Grant

After NSF selected the Harvard University proposal for funding in 1962 for the 1964-65 austral summer, the project was named *A Botanical Survey of West Antarctica* or *Bio-Harvard-Melchior.* Mack had also assigned the name "Operation Gooseflesh," the British form of goose pimples which causes little bumps on your skin when you suddenly get chilled.

We were informed officially and began to purchase, test, and evaluate equipment suitable for Antarctic research. Since we were going to be the first international SCUBA team diving for marine algae in Antarctica, there were almost no prior experiences or recommendations recorded. This meant that our project would undergo intense preparation and evaluation.

Two years prior to our departure to the Antarctic Peninsula, we began our year-round project in Rockport, Massachusetts. The purpose of the Rockport project was to test and evaluate diving techniques for observing, recording, and collecting algae, especially throughout the winter season. We made collecting trips nearly every two weeks for two years. During these diving and collecting trips, we also tested equipment and made modifications to improve our equipment to function in Antarctica. We made our neoprene hoods to cover our faces around the diving mask to keep us warmer. Diving in the winter season was difficult at first, but eventually became routine. We developed good collecting procedures and clear hand signals, and we became able to tolerate extremely cold diving conditions.

We knew we were going to be based at Melchior Base, an island on the Palmer Peninsula claimed by Argentina. The Argentine nautical charts were sent to our lab at Harvard. The charts, though good, were more suitable for ship navigation than for coastal diving operations. Eventually, we learned that since several nations claimed territory on the Antarctic Peninsula (Argentina, Chile, and the UK), they each had their own charts. Each of the three charts had named every island, channel, and shore land reference by a different name, so every reference point had three names. The only chart we obtained was from the Argentine Navy Hydrographic service.

Once we had our charts and decided on which names we would call our destination, we started work on the travel plans. Remember, no computers could do this research then. The initial plan was to fly from Boston to Miami; then to Panama; to Quito, Ecuador; to Lima, Peru; to Santiago, Chile; and finally, to Buenos Aires, Argentina. This would take us several days.

Before we left Cambridge, MA, we had to pack our diving and collecting equipment, box it up in reinforced wooden crates, and then secure them with steel strapping in case some cargo had to be air-dropped at Melchior. We had to get the equipment delivered to the US Naval Air Station Quonset Point at Davisville, Rhode Island. This base no longer exists, as it was closed in 1994. The cargo included lab supplies, microscopes, an air compressor, and SCUBA tanks. We decided to bring our personal air regulators as 'carry on' luggage. No TSA airport inspections existed in 1964.

Anyway, all of our diving and scientific supplies were packed up in 35 boxes and delivered in the early summer, with the expectation that we would collect it at Melchior later in the Fall.

Some of the wooden boxes packed with the equipment we needed for the expedition.

US Navy Deep Sea Diving School and Experimental Dive Unit

One aspect of the National Science Foundation's (NSF) funded Antarctic Diving Expedition was the requirement that each diver must successfully pass a rigid Navy physical. This entailed a very long day of physical testing and evaluation. There was a concern about one member of the Harvard crew in that he still had his appendix. NSF preferred that all members of the team had their appendix removed before going to Antarctica. I was 'OK' since I had my appendix removed at age 6.

The US Navy Diving School and the Experimental Dive Unit were located in the Navy Yard on the edge of the Anacostia River. The river, which feeds into the Potomac at Washington DC, has a nice brown color and appears to be filled with unidentified flotsam. Some of the flotsam included partially chewed food, feces, and bodies of dead rats. We now understood why we were required to get vaccinated for Tetanus, Typhoid,

The US Team at the Navy Diving School.

and four other waterborne diseases.

On another note, I and several others from our group were determined to be deaf and color blind. The deafness issue was initially scary, but luckily it was found to be a defect in the hearing test procedure. Failure of the deaf test would rule out any diving at all in the future, so the correction of the hearing test was most welcome. The color blindness test, which included identifying numbers and letters from a matrix of colored dots, was also a defective test since we could all distinguish the colors of objects in the room. A footnote to the color blindness test was the need to recognize the color of algae and their substrate for collecting purposes. The most prominent colors were blue-green, green, brown, and red algae.

All of us passed the rigid Navy Diving basic physical with more diving evaluations yet to come.

After numerous vaccinations because of the filthy Anacostia River, we began the two-week course. Our group included Jerry Koyman from the University of Hawaii who was an expert on deep diving seal physiology and was going to be diving through the ice at McMurdo Base, and two undergraduate students at Old Dominion College in Virginia. We started with an hour of physical exercise beginning at 6AM each day. Next, we spent a few hours in the 15-foot deep indoor pool on base.

Work in the pool included the usual activities found in our YMCA diving course. The basic elements included facemask removal underwater and ditch and retrieve exercises. No one told us about the blackened facemask exchange during a mask removal exercise. When a diver, in training, undertakes the ditching and retrieving exercises, their SCUBA tank, mask, and snorkel are removed from the diver and left on the pool bottom. The diver surfaces, takes a few breaths then swims to the bottom of the pool and retrieves his equipment. The instructor substitutes the divers mask with a facemask with a blackened faceplate, eliminating any visibility. Most Navy divers can cope with this kind of distraction and continue to gain confidence under difficult situations. After several days of pool testing we had the opportunity to receive training in the hardhat rig and the Jack Brown rig.

The hardhat diving suits were made up of a canvas rubber material. They were very large, like in a "one size fits all" style. The neck, armholes, and ankle holes were made of stretchy rubber and were tight around the openings. Underneath each diving suit we wore waffle knit long john underwear. Once the diving suit was on, a fifty-pound weight belt was attached to the diver. We did not use inflatable life vests since we could inflate the entire suit by simply turning an air hose valve located on our stomach.

Next, we stepped into weighted shoes of about 20 pounds each. Finally, two people would attach the metal helmet to a metal collar on the canvas diving suit. I noticed that my helmet was covered with numerous bumps and large dents; it made me wonder what I was getting into. The helmet had round glass portholes on the front and the same on each side. The glass was protected by metal bars, much like a baseball catcher's mask.

Once in the 15-foot deep test tank, I felt water entering the canvas suit as I slowly sunk to the bottom. The incoming water accumulated in the leg portion of the suit. To remove the water, I had to stand on my head and vent the water out by using my chin to open an exhaust valve. The valve which extended into the helmet was reached by turning my head to the right side and pushing it with my cheek. I tried to evacuate the water from my suit several times and eventually found it fairly easy.

During the second week, we trained further on hardhat and got a chance to dive in the muddy Anacostia. One exercise, using SCUBA gear, was a bottom search pattern in case we had to search for a body with no underwater visibility. The course meant diving to about 30 feet deep in the muddy Anacostia with a hundred feet of line weighted with 20 pounds on each end. With zero visibility, we swam to the end of the line then we advanced the weighted line about 10 feet and swam the line in reverse. By moving the line at the both ends we created a zigzag search pattern. The river experience was something else. Fortunately, we never found a body.

Next on the agenda was regulator maintenance and repair. I enjoyed this because it could be very useful if our regulators froze up in Antarctica's freezing water. I enjoyed learning how to take a U.S. Divers regulator apart, repair any problems, and put it back together again. We were not going to be using Hard Hat gear in Antarctica, so this lesson was not really important for our safety.

The first days of the course included registration and more physical evaluations. Next was familiarizing ourselves with some of the principles of physics and medicine, as well as the gas laws such as Boyle's law, Henry's Law, Dalton's Law, Charles's Law, and Archimedes' principle. We also covered causes, effects, and treatments for topics like air embolism, pneumothorax, Nitrogen narcosis, and hypothermia. Repetitive dive schedules were also introduced.

SCUBA training in the pool lasted for about a week and included orientation and maintenance of equipment, different methods of entry into the water, clearing the face mask, buddy breathing, ditching and retrieving scuba tank and regulator, use of a flotation vest, and a final harassment in the water.

Practical training took place in the Anacostia River, a feeder to the Potomac, close to the D.C. Navy Yard. The Anacostia River was very dirty and polluted. Underwater visibility close to the surface was less than one foot. In deeper water, 20 to 30 feet, the underwater visibility was zero. All training swims underwater were made in buddy teams of 2 or 3 divers. They were secured to each other by means of a 6- to 10-foot-long buddy line attached to each diver.

The section on equipment repair and maintenance was very interesting and well taught. We were all able to dissect our own our own regulators, change any defective parts, and reassemble them in a short time frame. We also learned how to use, maintain, and fill our scuba tanks using a small compressor.

We even spent some time familiarizing ourselves with hardhat rigs and Jack Brown full-face masks. Hand signals and standard Navy tending signals were also included. We studied Decompression Tables and learned how to plot all aspects of a dive. Then we all qualified for a 130-foot chamber dive and a comprehensive final exam. All of our team qualified with great scores on the exam and were deemed qualified to dive in Antarctica.

Chapter 2
The Journey

"Antarctica has this mythic weight. It resides in the collective unconscious of so many people, and it makes this huge impact, just like outer space. It's like going to the moon."
Jon Krakauer

ON OCTOBER 1, 1964, Richard Waterhouse, an additional technical diver on our team, and I began the actual traveling part of our Antarctic trip. All of our cargo and supplies from Harvard were delivered already to the Antarctic Program Office in Davisville, Rhode Island. From Davisville the cargo would be delivered to Buenos Aires and transferred to the icebreaker *San Martin*. We boarded the plane at Logan Airport in Boston bound for our first stop at the University of Miami. Once in Miami we were met by John Walsh, a Harvard graduate, who was now a graduate student in Marine Biology at University of Miami. We spent the day touring U Miami marine lab and then we boarded a plane for Panama City. We found a cheap hotel, had some dinner and went to bed.

The next day, we took a taxi tour of the city; the barrio or poorest section, the Panama Canal, the Miraflores locks, and the surrounding jungle. I never realized the canal project was so immense. We even observed large cargo and tourist ships passing through the locks.

The next day, October 3, we arrived in Lima, Peru where we stopped at a luxury hotel surrounded by houses constructed of mud and straw. We made contact with Father Lawler, a Maryknoll priest and a New Bedford friend of my fiancé Katy's parents (small world). On the 4th we took a train to visit the Andes town of La Aroyo at an elevation of 13,000 feet, I quickly noticed the reduction in ambient oxygen. We felt like we were drunk, kind of dizzy and wobbley on our feet. It was like getting off a sailboat after a windy day in Buzzards Bay. I met a Canadian biologist and we tested a Pisco sour, a traditional drink for high altitude. The Pisco sour made me drowsy but did not alleviate my response to the high altitude. Back at the hotel I very quickly fell asleep.

On October 5th, we took an early bus over more mountains, to Huancayo at an elevation of 16,000 feet. The bus was really crowded with all luggage piled on top of the beat-up bus roof. We looked like a band of gypsies. The Indian woman sitting beside me kept pulling out her breast and nursing her child. She also changed the baby's wrapping several times by removing the soiled area and relocating it away from the baby's skin. This was quite an experience for me.

After four or five long hours and a few pit stops we finally arrived at the state owned Hotel Tourista. We gathered our luggage (several brown army duffle bags) and set out for the hotel across the street, about 50 yards away. The small amount of oxygen at 16,000 feet left us gasping for air, so we took up the luggage assistance offered by some young boys. We could not believe how the lack of oxygen affected us. Another observation: all of the local Huancayo people were very short in stature, mostly fat and by their clothing and derby hats very poor. We also saw that many of the locals had large swellings in their cheek. We were told that the swellings, about the size of a marble, were from chewed up coca leaves which when exposed to enzymes in the mouth released cocaine. The cocaine dulled the senses and warded off misery from hunger. Coca was easily available from the taxi drivers and other local sources. I wanted to give the coca leaves a try, but was afraid I could get robbed so I did not. We went to bed early after eating a great local dinner, which included some indescribable meat and several types of unusual looking potatoes. I have since learned that Huancayo is the location of the largest collection of hybrid varieties of potatoes in the world. Early the next day we did a little souvenir shopping and boarded a train that traveled on a switchback train route back to Lima. The trip only took 10 to 11 hours and we ate a hot meal on the train. In Lima, we stayed at the Hotel Alcazar where we met several Peace Corp volunteers. We talked with them about Peru until early in the morning. One of the Peace Corp volunteers showed us the open market where I bought hand-made shawls for Katy and a good friend, Jane Shea.

The next day we flew to Santiago, Chile and arrived at lunchtime where I had my first 'empanada'. It was delicious. The empanada was hamburg wrapped in piecrust dough and deep-fried. They were cheap and good. Eventually we took an overnight train traveling 600 miles south of Santiago to an extinct volcano crater called Llaima, which still had skiable snow. We hoped we could experience a few days of skiing and we had the time to do it. Richard and I shared a train compartment with a tiny toilet and washbasin; we slept well and arrived at Lima before noon. The next morning, we found a bus which took us to a mountain trail leading to the Refugio (rugged ski lodge). After an hour and a half hiking uphill in the snow carrying our heavy luggage bags we finally got to the Refugio. This place was interesting, small spartan rooms with cots, mattress, small table, coat hook and a sheet for the door. The common language here was mostly German. This was actually OK because Waterhouse spoke fluent German. The manager of the inn was Fritz Karstens, ex Gestapo officer and nasty to the core. He made no secret of the fact that he hated Americans. He also had one of the largest short wave radio systems in South America, which he used to talk to family and friends in Germany.

For the next few days Richard and I used borrowed ancient skis to ski on an extremely steep slope with a most primitive ski lift. To ride the ski lift we had to rent a canvas belt, for 50 cents a day, with a special clamp on the end that hooked both ends together and clamped on to a moving steel cable. If you fell while on the lift, the clamp would come undone and you would disconnect from the moving steel cable. If lucky, you could get out of the way of the next person in line. The skiing was great.

We returned by train to Santiago and arranged a plane flight from there to Buenos Aires. In Buenos Aires, we headed to Darsena (Dock) A to find the icebreaker, *San Martin*. We soon found the ship and were welcomed, fed and told that the ship's departure for Antarctica would be delayed for a week so that radar equipment could be installed and tested. Since we had only minimal cash we were invited to stay on the ship. Dr Lamb arrived the next day so we sat in the ship's lounge and talked for several hours. The three of us were really anxious to set foot on the great white continent.

Ice Breaker San Martin in dry dock, Buenos Aires.

On October 27th, Lamb, Waterhouse and I were finally settled in our new cabins on board the icebreaker, Rhompehielo *San Martin*, where we were to live for the next couple of weeks. I met the ship's resident dentist who fixed my tooth where a filling had fallen out. Lucky for me he spoke English, far better than I spoke Spanish, which down here in Buenos Aires sounds a lot like Italian I heard in Boston's North End. Later in the day I met the ship's doctor who gave me some pills for diarrhea that I'd had for two days.

Dr. Lamb arrived late that night, and the next afternoon he went to the "aeropuerto" to meet Dr. René Delépine, a research professor from the University of Paris (Sorbonne). Delépine was an expert on Antarctic Marine algal taxonomy, the science of categorizing, identifying and naming algae. Since he was not a SCUBA diver, he was really excited to join our expedition so he could work on algae that we collected from the shore and at greater depths.

Delépine arrived at the *San Martin* with a severe cold that he brought from Paris. He was totally exhausted from his 36-hour flight from Paris to Buenos Aires via the Canary Islands and Dakar, Africa. He felt much better after a quick meal and 20 hours of sleep.

The next morning the ship's dentist took me into Buenos Aires to visit the "Bulls Balls" (another name for the black market) where I could get a good exchange rate for dollars to pesos. We found a store where I could purchase a salida de banõ (bathrobe). I also bought shower clogs, soap, deodorant, etc. for 5 months' time.

That afternoon the *General San Martin* left port for two days to test their new electronic equipment (RADAR). I finally met Delépine who seemed very pleasant. Lamb spoke fluent French, but Waterhouse and I did not. Conversations were difficult, but we all got along well.

After a great dinner in the officers' dining area I made some observations on the furniture adaptations. First, there were some small lengths of chain, which attached the arms of the comfortable chairs to the table. This was to prevent the chairs from sliding or tipping if the ship was pitching or rolling too much. There was also a configuration that popped up at each table placement, called fiddlers. They were a slightly elevated framework or table with cut out holes to fit plates, a glass, cup and silverware. The purpose of these was to prevent dishes of food from sliding off the table when the seas got rough. I began to wonder: how rough?

The ship was now back at its berth at Darsena A in Buenos Aires harbor by Friday. Interestingly, Buenos Aires is located 160 miles upstream in the Rio de la Plata (Silver River) from the open sea. We were next taken to the Antarctic Institute where we met Admiral Panzarini, the director, who had built the Argentine base called Almirante Brown before he retired from the Argentine Navy. We had lunch at the swank Argentine Automobile Association Club, where we met a few more scientists who had also been invited to lunch. Later in the day we retrieved some air freight (microscopes) for Delépine. Delépine was scheduled to fly to Puerto Deseado the next day, and then he planned to meet us when the *San Martin* arrived at Ushuaia. I got the idea that Delépine didn't particularly relish time on a ship.

We continued to live on ship because the length of the delays was not clearly communicated to our group. Besides I did not have enough money left for a week or two in a hotel. The officers' menu was great: buns with butter and some kind of jam worked fine for breakfast. There was steak for lunch and dinner. I had never had so much steak in my life and I loved it. Remember, I came from Somerville, MA and people from my town didn't get to eat steak very often, and certainly not twice a day. I neglected to mention the salad, which I had to stop eating because I was convinced it was the home of numerous ameboid pathogens, which caused Americans to have diarrhea. So I kept well away from salad.

Lieutenant Cypriano Olivera, our next-door neighbor on the ship, was an officer in charge of 'guard duty' for that night. He spoke some English and seemed to be a lot of fun, much like my best friend Kaz. He taught me "rasca su bolas" (scratch your balls), a common phrase in the Argentine navy. He and his fiancé planned to get married at the same time as Katy and I. I also learned honeymoon in Spanish: "Luna de Miel."

On Monday, Lamb, Waterhouse and I went to the American Embassy to see if it was possible to purchase some liquor and cigarettes. I got several cartons (I smoked then) and an assortment of Drambuie, bourbon and cognac. We actually purchased a fair amount of liquor because we had to be prepared for a long time on the ice. We thought that liquor might also be good for trading with others, and it was tax free and very cheap.

I wrote a letter to Katy but the postal service was on strike for the past two weeks and still remained on strike and I was unable to mail my letter. After dinner, we went to Augustine's (the dentist) house to have a drink of his special whiskey. It was smooth and good and my first introduction to Argentine moonshine. He completed work on my bad tooth earlier and it felt great. It would be a big problem trying to dive in Antarctica with an exposed tooth nerve. Lucky me!

The sailors painted the ship the usual battleship gray. The Argentine navy somehow got the ship from the German navy, part of the spoils of WW II. I was thoroughly surprised at how nice the ship was in the officers' section, and not as great in the enlisted sailors' quarters.

We finally took our dirty clothes to the ship's laundry. It's been over two weeks without laundry service. In the evening, we again went to Augustine's house to meet his parents and his wife. His father gave me two bottles of his special wine and told me I was his adopted son. We had a great time there and realized how caring the Argentine people were toward us.

On November 4th, Johnson won the US election for president. This was good news to share as we entered the ship's departure dinner. I met the ship's priest who gave me a holy medal for Katy.

Our pre-Antarctic stay on the icebreaker *San Martin* was valuable in that we were able to observe all the preparation steps for this trip in advance. The installation of radar was a surprise to Richard and me. We automatically assumed that radar was already a part of the ship. It was no surprise that sailors were grinding rust spots off the hull and repainting those spots battleship gray. On one occasion a cargo net loaded with bags of sugar was lifted by the ship's crane from the dock, over the sailors painting the hull to the cargo hold. Suddenly some of the cloth bags of sugar burst open, falling directly onto the sailor painters and onto the freshly painted hull. No big deal, without even slowing down the painters dipped their paint rollers and painted right over the sugar newly stuck to the fresh gray paint. Nice sugar coated ship.

Ship loading continued and we saw truckloads of food on pallets, which were placed in the cargo hold. There were also over 100 half cows, already slaughtered, ready for the cold storage. With all the food and supplies loaded we were almost ready to go.

A couple days later, we finally met Lieutenant Eduardo Marcello Cueli, who was to be our base Commander. He took me to a music store and helped me purchase a guitar. He himself was an accomplished classical guitarist and I was a folk musician.

By the 9th, the ship was loaded with equipment, supplies, food and mail. We were ready to embark following a departure ceremony, blessing, speeches, and many goodbyes from friends we had met at the University. We set sail at 1540 and began our 150-mile trip on the Rio de la Plata River, followed by another 400 miles to Ushuaia. There we would take on fuel and water and continue on to Antarctica.

The next day we were cruising the Rio de la Plata, with smooth sailing thus far. We found out at dinner that all the butane bottles containing gas to run the room heating at Melchior had not been filled before they were deposited on the ship. This meant that our rooms, on the base, would not have much heat at night. I wasn't sure how this problem would be solved.

We sailed into the Beagle Channel, made famous by Darwin over 100 years ago, and the ship docked at Ushuaia, Tierra del Fuego, Argentina, the southernmost city in the world. We were able to leave the ship to see the town. When we were leaving the ship we immediately felt the constant wind at about 30 miles per hour, and saw some snow blowing in the wind—this is summer?

Days on the ship could be boring, here are some interesting excerpts from my journal:

18 November

"Today we left Ushuaia the weather was cloudy with fog and intermittent snow. We are now in the Beagle Channel headed for

the Drake Passage. For now we are feeling wind from the southeast and the ship is rolling constantly. The officers here claim the roll of the boat is so great (60°) that you can walk on the walls."

19 November

"I woke up early at 0600 and had some coffee and rolls. After I went to the bridge and found that we had anchored behind Picton Island at the mouth of the Beagle Channel. A Chilean submarine surfaced astern of us and three Chilean cruisers circled about a half mile away. The weather is cloudy, hazy with occasional rain and snow. Miserable weather, but better than yesterday. The ship was facing SE and into the wind, 40 mph and now pitching to and fro mildly."

San Martin in the Drake Passage on a good day.

20 November

"We are underway at 0845 this morning. We were about to enter the Drake Passage 'en route' to the Great White Continent. Metal hatches have been ordered secure on our portholes. Just after leaving Picton the crew passed out 6 foot long and 4 inches wide straps to help keep us from falling out of our bunks. I expect the next 36 hours to be very rough. After 40 hours in a raging sea the weather began to settle down."

21 November

> *"We can see icebergs, penguins, and coastal regions. The San Martin is approaching Deception where we expect to stay for two or three days before proceeding to Melchior."*

Deception

Our first stop on the Antarctic Peninsula was Deception Island. This is geologically considered an "extinct" volcanic crater. We were anxious to see what an Argentine naval base looked like. There are three bases located within the ocean filled crater, Argentine, Chilean and British. As a result, we needed to be familiar with at least two languages since we would be at Deception for a length of time.

This was a dull weather day 'en route' to Deception with an anticipated arrival of around 2400. We began to see lots of icebergs and plenty of penguins. The *San Martin* slowly entered the narrow, shallow opening to the water filled top of this 'extinct' volcanic crater. Once the ship was inside the crater, it moved close to the Argentine research station and set anchor. The holding bottom was good here and the spot was protected from severe winds by the high mountains, which make up and surround the crater.

We had now reached the point where there was no distinction between day and night. Night was gone and daylight prevailed for 24 hours at a time. There was also a heavy covering of fog that surrounded everything.

When we awoke in the morning we were securely anchored inside the extinct, seawater filled volcanic crater, caldera in Spanish, of Deception Island. Heavy fog and falling snow made visibility difficult. A launch from the ship had already been deployed toward the Argentine base. This was the first ship to arrive here after the winter season. Year-round occupants of this facility were very anxious to receive mail from the ship. Fresh fruit from the *San Martin* was also on that first launch. My journal from that day reads: "A shift in the wind caused pack ice to drift and surround the anchored ship rendering any further cargo exchange impossible. Those sailors who went to the base on the first launch were blocked from returning to the ship, unable to move the launch through the ice. No problem for the crew, they could stay warm at the base."

Unloading equipment at Deception during a whiteout.

Later that day, the *San Martin* changed its anchorage into some deeper water and we were able to hitch a launch ride to check out the base. When we landed at the edge of the shoreline ice, the fog and snow came back and we experienced our first 'whiteout'. A whiteout occurs when a base or any solid object disappears from view because the fog blends with the snow-covered background and you become blinded and disoriented. Whiteouts also occur, as in our case, when the ship was anchored and suddenly within minutes the ship became enveloped in a dense fog and disappeared from view. I happened to have a camera on hand when I landed on the edge of the Deception shore ice. The *San Martin* disappeared from view and the launch could not return to the ship. At the same time Deception Base, surrounded by fog, also disappeared. I was actually able to photograph the *San Martin* as it disappeared before my very eyes.

Indeed, the weather could change in a minute. I could only begin to imagine the potential problems diving with an unanticipated "whiteout" present.

A visible San Martin in the top photo and a few seconds later the ship is not visible in the bottom photo.

Monday was a great day with lots of sun, no wind, and a flat calm sea in the caldera. When I made another visit to the base I met one of the officers in charge of loading and unloading cargo from the small launch that moved back and forth from the ship to the edge of the ice. When the launch landed close to the edge of the ice a large plank was set up from the launch bow to the ice. This allowed cargo to be off loaded by sliding boxes down the plank from the launch to the ice. The concern was that none of the cargo slides off the plank into the sea. The officer had gone onto the launch to give some orders; when he came back down the plank, he lost his balance and slipped into the freezing water.

The sailors were able to grab the officer, pull him out of the water, and quickly remove his clothes. The sailors each contributed a piece of their clothing to warm the officer who was at the risk of hypothermia in -2° seawater. He was quickly brought to the nearby base, warmed up, and returned to duty in less than an hour. I was really impressed at how quickly the sailors dealt with a potentially serious condition.

Delépine and I spent the rest of the day walking along the shore collecting some algae on the beach and some lichens along the edge of the shoreline. The beach was composed of fine, almost glassy, black volcanic sand. In some spots columns of steam were emerging from the sand. These are called 'fumaroles'. If our hands were cold, we could warm them up by sticking them into the black sand. I did not fully realize that these fumaroles were indicative of some degree of volcanic activity from far below the surface.

This is the first time we really saw lots of penguins too. My journal reads, "We heard lots of noise coming from a small hill on the side of the caldera. Upon inspection, we observed a massive penguin rookery. There were thousands of penguins sitting around on stone piles. I was not sure what species since these were the first penguins I had ever seen. I will say that penguins are noisy and smell horrible."

On our way back to the base we observed some large chunks of ice washed up by the tide on the black sand beach. The upper 12 inches of the ice chunks had several species of algae frozen into the ice. This seemed to be a mechanism of biological transport for moving species of algae from one place to another. How far these algae could travel remains completely unknown.

We next visited the Deception base where we met other Argentine researchers and naval support crewmembers. Lamb observed in his journal that in the main lounge of the base three small flags of equal size were placed side by side: Argentine, British and Chilean. I saw this as a pleasing demonstration of the spirit of mutual tolerance, which had emerged in recent years since the signing of the Antarctic Treaty.

The Antarctic Treaty of December 1959 was signed by the governments of Argentina, Australia, Belgium, Chile, The French Republic, Japan, New Zealand, Norway, The Union of South Africa, The Union of Soviet Socialist Republic (now Russia), Great Britain and the United States of America. The treaty is an international agreement that assures the use of Antarctica for peaceful and scientific purposes only. It also includes 'International Cooperation' and scientific exchange of personnel. Our group exemplified the goals of the Antarctic Treaty: Lamb, Great Britain; Zimmerman, Switzerland; Delépine, France; Bellisio, Argentina; Waterhouse and me from the USA.

In the spirit of international cooperation we learned the scientists and crew members from the French base, the Chileans, and the Argentine base gathered frequently, weather permitting, to party. The Argentines provided 'carne asado' or good quality beef; the Chileans provided quality red wine and the French gave out 'Galois' cigars.

The next day, at about 0800, we hauled anchor and crept through the narrow, shallow entrance to the crater. In this channel we saw the remains of two ships that had smashed up on the rocks many years ago. They were heavily oxidized and covered with rust. We were soon back at sea headed to Melchior Island. After 12 hours of cruising between numerous small islands and several Fiords, we finally saw the location of Melchior.

We arrived late in the afternoon at Melchior Island and set anchor on the north side of Isla Observatorio (Gamma Island). I should point out here that several nautical charts are used to designate locations and channels on the Palmer, which is now called the Antarctic

Peninsula. The names of geographic places, islands and beaches are labeled differently on each of the charts. The charts are produced by hydrographic services of Argentina, Chile and the UK. Therefore, any given island along the peninsula will automatically have three different names.

It is very confusing to navigate among these islands, it is also extremely isolated. But we finally arrived to find Melchior Base covered completely by snow. We set anchor and began the process of lowering the first boat to visit the island.

Chapter 3
The Melchior Base

"Southwards, a magnificent Alpine country, illuminated by the rising sun, rose slowly from the sea; there were mighty fells with snowy crowns and with sharp, uncovered teeth, around the valleys through which enormous, broad rivers of ice came flowing to the sea."
J. Gunnar Andersson, South Georgia Island, 1902

AFTER A 12-HOUR CRUISE from Deception, the *San Martin* and its research team arrived at Melchior Island. We knew that Melchior Base, also known as Destacamento Melchior, was located on a small peninsula that had two arms extending in a northerly direction. In addition to the main house there were some smaller sheds and several tall radio antennas; an emergency house (dog food and emergency food and supplies were stored here); a jetty with a small crane for unloading food and equipment; a small cold storage house; a workshop;

50

and fuel storage hut. Since a fire in the main house could be potentially devastating, the smaller buildings were arranged at a reasonable distance from the main building.

The ship anchored off the north side of the island, but we couldn't see the base, just a mountain of snow. We did observe several radio antennae projecting out of the snow and a small emergency house close to the ocean.

Many Antarctic scientific bases were surrounded by small, woodshed size buildings or shelters where researchers could shelter during the occasional whiteout. Over many years these shelters became known as emergency houses. Often, due to shortages of space in the main base, the emergency houses were used to store tons of dried dog food. Dogs have not been used on any Antarctic base for many years, however if a party were stranded they would not have starved. Although the dogs were no longer used to move equipment around on Antarctic islands their food supplies still remained. We tasted some and it was terrible. Later that day my voice developed a slight growl.

There was also a small, dark structure lying on the top of a snow ridge, which was actually the top of the chimney to the base building hidden

The small black line in the center is the top of the base chimney.

below the snow. I was surprised at how desolate this location seemed. No doors, windows, or structures were recognizable due to the heavy snow cover accumulated by many years of being unoccupied. The thought that the base could be filled with snow and ice passed through my mind. Maybe it had collapsed and only the chimney was left. A few other small buildings were buried without a trace, such as the Argentine Navy crew house.

A cargo barge was deployed off the icebreaker to carry some of the scientists and naval crew from the ship to a small stone dock with stone stairs extending into the water. To get off the barge that ferried us to the base emergency house, we had to chop the ice accumulated on the stone steps leading to the small building above.

Many of the officers, sailors, and scientists began the task of digging out the base. Five small shovels were passed around from person to person, each of whom took a turn for about a fifteen-minute digging shift. After about an hour the crew cleared around the chimney and reached the peak of the roof.

Digging around the chimney

Eventually we stopped digging by the chimney because we realized we were digging in the wrong place. After some discussion, we moved the digging site to a place estimated to be near the front door of the base. After 20 feet of snow removal we saw the top frame of the doorway and were able to clear a pathway to the door. The door was unlocked of course. Nothing inside was worth stealing. The doorway was on the corner of the building facing the small harbor, also called a caleta, which opened to the northeast.

Everyone wanted to be first to enter the base. Mack pointed out that, "This is like breaking into a sealed burial vault." Someone in our small group had a flashlight and upon entering the hallway found some candles, which quickly lit the way. The first thing we noticed was that the linoleum floor was soggy and spongy underfoot. All the rooms were in reasonable condition, but by our standards, the base overall was in tough shape—buried in the snow, no heat, no cooking facilities, no lights,

Tunnel to front door after two weeks of digging.

53

no generators, and no toilets or showers. Inhospitable to say the least.

I learned later that Antarctic tradition required that a surprise and a welcome note be left on the table for the next team to occupy the base. Our surprise was a large supply of Cinzano vermouth, wine, loaves of salami, and some crackers. This was a nice treat and time out for those involved in the snow shoveling.

While we were shoveling snow at the island base, the ship's diver got drunk and had an appendicitis attack. He had severe stomach pains that were quickly diagnosed by the ship's doctor. The appendectomy was undertaken in the ship's infirmary, but not without incident. The diver had a severe reaction to the local anesthesia. After the surgery, he was resting comfortably. The patient was in good health and walking around the ship within 24 hours.

The next day no boats were able to bring men or cargo from the ship to the base because the caleta was filled with icebergs that completely clogged access to the landing dock. Later in the day, the launch broke down due to a broken starter. The ship's mechanics were able to fix the launch so it was ready to load for the next chance to get in the harbor. Seals could be seen swimming around the ship as well as the launch. The seals had never seen ships, boats, and people before, so they had no fear. There were about 20 seals that seemed fearless and inquisitive.

It seemed like the rotation of the earth forgot about night as there was no darkness at night. The Captain's launch broke down on its return to the ship from Melchior base. It drifted easterly behind some large icebergs. No other launches were in running condition, so the ship (*General San Martin*) lifted anchor and went to the rescue before the launch was caught between the large icebergs and crushed or overturned. The launch and crewmembers were rescued, but the icebreaker ran aground. We were able to reverse, back off, and get back into deeper water. Lt. Doncel said in his journal that, "in another 45 minutes the launch and men could have been lost forever." This event reminded me that you always need a plan B to cope with the unexpected which could be mechanical breakdowns or rapid changes in the weather.

Since the base was in such poor condition, that is no heat, light, toilet, water, etc., Lamb and Captain Bustamante decided to leave some sailors at Melchior to shovel the heavy snow from the roof so it wouldn't be crushed when the furnace started working. The estimated depth of the snow from the rooftop to the ground level was about 25 feet. Our scientific crew would continue to live on the *San Martin* for the next 5 to 6 days while the ship resupplied another Argentine base called Almirante Brown. A crew had wintered over at Almirante Brown and was anxiously awaiting mail and fresh fruit.

Almirante Brown was a warm and friendly base. We found that visiting there was especially comfortable and cozy. They had a great docking area and it was easy to offload a new crew, food, and supplies. Most of the crew there were meteorology people assigned by the Argentine navy.

While we were at Almirante Brown we decided to check out the intertidal area around the dock looking for algae. We still had not unpacked our diving suits and equipment from the boxes, which had been unloaded at Melchior. The shoreline was devoid of algae.

There was no beach area to collect algae. The shoreline consisted of large volcanic boulders and volcanic substrate. Lamb and I made an effort to snatch some specimens using a knife tied to a stick. The water was exceptionally clear, but I could not pry off and collect any complete specimens of algae, only fragments that didn't count as research.

After 2 days, the *San Martin* cruised back to Melchior to see how things were going. The sailors had devised a method of snow removal which involved cutting the snow into blocks and passing the blocks along a line of men, the last of whom pitched the block into the sea. The roof of the base was now more than half exposed and a tunnel had been excavated from the main dock to the front door of the building. The sides of the tunnel were about 20 feet high. Cargo could be offloaded from the launch, placed on an akio sled and hauled about 100 feet up to the front door. Extra heavy items, such as generators, were hauled into the hallway and right into the kitchen area. Kind of messy at the end of the day.

The Crew

There were 13 people that lived on the base, including scientists, divers, and support crew from three countries—Argentina, France, and the United States. The support crew included:

- Sr. Gines Garcia—Support crew
- Sr. Aldo Gubolin—Support crew
- Sr. Juan Morales—Support crew
- Sr. Domingo Yunes—Argentine Navy, radio operator

The scientists included:

- Dr. Norberto B. Bellisio
- Teniente (Lt.) Eduardo M. Cueli—Head of Base
- Dr. René Delépine—Scientist
- Mr. Richard A. Fralick—Technical assistant, head diver
- Dr. Ivan Mackenzie Lamb—Chief scientist, diver
- Sr. Hipolito Ojeda—Argentine Navy, cook
- Mr. Richard E. Waterhouse—Technical assistant, diver
- Dr. Martin H. Zimmermann—Scientist, diver

Here we all are together.

Dr. Norberto B. Bellisio, an Ichthyologist Specialist in Antarctic fishes, Museum of Natural History, Buenos Aires, Argentina, worked on collections and identification of Antarctic fishes. He was a very quiet researcher with a superb sense of humor. He studied age-related development of fish livers.

Teniente (Lt.) Eduardo M. Cueli was an administrative officer for the Argentine Navy, Hydrographic Service. He was the Chief of Station for Destacamento Melchior. In that capacity, he was in charge of opening and maintaining Melchior Base and the supervision of all aspects of running an Antarctic research base. He was also a skilled navigator and a SCUBA diver.

Dr. René Delépine was from the Museum of National History, Sorbonne, Paris, France, and a Seasonal Antarctic algal researcher and botanist. He joined our group late in the development of this expedition, when he met Mack and was invited to come along and join the survey.

Richard A. Fralick (me), was the Chief Diver and Divemaster for the expedition. I was a SCUBA instructor at the Cambridge YMCA and had worked at the Harvard University, Farlow Herbarium as a diver and technical laboratory assistant for three years. I also worked with the diving logistics and research on algae for the Harvard Rockport algal study for two years. I attended the U.S. Navy Diving School with Lamb, Zimmermann, and Waterhouse.

Dr. Ivan Mackenzie Lamb, aka 'Mack', Chief Scientist, Botanical Survey of West Antarctica, received his Doctoral degree from Edinburgh University. During World War II, he participated in an Antarctic Expedition (Operation Tabarin). He was the recipient of the U.S. and British Polar medals. He served as Professor of Cryptogamic Botany at Tucuman University in Argentina. He also conducted research at the Canadian National Museum and in 1954 he became director of the Farlow Herbarium at Harvard. He retired from Harvard in 1972. His research covered terrestrial and marine Lichens and Algae. He was a trained SCUBA diver and was Chief Scientist for the 1964-1965 Harvard Antarctic Expedition.

Hipolito Ojeda was an Argentine Navy cook. He was a good cook and enjoyed the company of everyone at the base. Domingo Yunes was the Argentine Navy Radio operator who kept us in touch with the rest of the world—sometimes.

Richard E. Waterhouse provided logistic support for delivery of our cargo to Antarctica and its return to the U.S. He had several years' experience SCUBA diving and was involved in the technical diving for the Harvard Rockport Study. Richard was fluent in Spanish and German and also familiar with boats.

Dr. Martin H. Zimmermann received his Ph.D. from the Swiss Federal Institute of Technology in 1953 in the field of Plant Physiology. Martin was Professor of Physiology at Harvard University and the Director of the Harvard Forest research facility in Petersham, Massachusetts. He conducted his research at the Harvard Forest, Halibut Point in Rockport, Massachusetts and in Antarctica. He was an accomplished musician and SCUBA diver. He was a Fellow of the American Academy of Arts and Sciences.

The Living Working Hut

The interior of the main hut was laid out according to the sketch that also identifies the cabin assignments, kitchen, dining room, lab space, radio room, bathroom and scientific supply room.

The external dimensions of the main living, laboratory and storage building were 85 feet long (28.5 meters) by a width of 25 feet (8.2 meters). The hallway was about 6 feet wide (2 meters) and extended along the main length of the middle of the building. Each of our individual rooms was about 10 feet long and 8 feet wide. The main laboratory consisted of the equivalent of two bedrooms.

Upon entering the wide front door there was a large dining room to the left with a long table and chairs on each side. This also served as a living/ reading/ quiet room during hours between meals. We also held our daily planning sessions here. On the back wall was Zimmermann's room, then my room. The laboratory where algae were sectioned and studied microscopically was next, followed by Lamb's room, then the laundry,

Main Building Floor Plan

Front door

Dining & Living Room	
	Zimmermann
Guests	Fralick
Cueli	LAB
Delépine	
Bellisio	Lamb
	Toilet Shower Laundry
Waterhouse	Kitchen
Radio Room	
My Lab	Storage

Back door

toilet, and shower room. Next came the kitchen, the domain of the cook, Hipolito aka Polito. In the kitchen, in addition to the stove and refrigerator, was the caldera or snow melter, through which melted snow was hand pumped to the cistern in the attic above. On the other side of the hallway, to the right when you entered the building, were rooms for Cueli, Delépine, Bellisio, Waterhouse, and a guest room; then the radio room, a storage room, and a large lab across the end of the building that I used. Each room had a window with no view, because the sides of the building were covered with snow.

This was my room.

After a week, a pathway was excavated of snow from the crane jetty to the front door. The pathway and doorway were wide enough to allow wide power generators and a heating furnace to pass through. Another pathway got shoveled out from the front door to the emergency house where we kept our diving compressor, air tanks and other diving equipment. A small stone platform and several stairs on the ocean side of the emergency house allowed us easy access to our two 12-foot wooden diving, rowboats. This platform also allowed divers to easily enter and exit the water of the caleta.

Attached to the kitchen, beside the back door, was a snow melt tank (caldera) with an outdoor connecting chute. Typically, snow was gathered daily and shoveled into the caldera in the kitchen. When the snow melted, it was pumped manually into a 300-gallon cistern located in the attic over the kitchen. When filled, the cistern held the weight of the 300 gallons or some 2400 pounds. This amounted to one ton plus 400 pounds as well as the weight of the cistern itself. We would have been flooded out if it ever broke. We now understood why it was critical to remove tons of snow from the roof.

Water was available only to the bathroom and the kitchen by gravity. Showers were limited by special rules and toilet flushing was restricted only for the important stuff.

Daily Routine

Once we got settled into something of a routine, the most critical factor governing the course of our field and laboratory research was the fact that the day never seemed to come to an end. During the austral summer season daylight is continuous for 24 hours at a time. That is, NO night time and NO darkness to end the day and transition into night.

Our Circadian Rhythms advanced and we found ourselves talking and sharing music until 3 or 4 AM. Bedtime was usually between these hours. Several of us initially had difficulty sleeping because the constant daylight hours altered those rhythms. A partial solution was to cover the windows from the inside with dark paper and exclude the penetration of sunlight through the snow pack and into our bedrooms. It worked. By darkening the room, our brains interpreted this as night and we were able to sleep. Most of us were up by 9 or 10 AM.

We had fresh baked rolls and guava jam every day for breakfast, along with the worst coffee in the world. Tea was also available. The crew mostly drank matté out of special gourd cups with silver straws that had a rounded end with small holes to sip the matté without getting the leaves in our mouths. It was brewed from dried yerba matté leaves. Matté was a traditional drink in South America, especially Argentina.

People were able to do whatever work needed to be done to get the day rolling. In the morning, Waterhouse and I typically filled air tanks and prepared diving suits, weight belts and other equipment for an afternoon dive. Other small tasks included cleaning up labs, sorting and pressing algae, and shoveling snow around the base which we melted for water.

Lunch time was usually between noon and 1 PM. At this time the scientific party and Lt. Cueli, the base commander, sat together and discussed plans for afternoon dives. The lunches were always good—soup, excellent beef, potatoes or rice, and sometimes canned vegetables. Argentina was known worldwide for its beef quality. In fact, when the icebreaker *San Martin* was off loading our food supplies, a side of beef was allocated for each of the crew members.

All the meals were prepared by our sailor cook, Paulito, who did not speak Spanish (the language of our expedition), but an obscure mountain dialect called "Guaranee." He came down from the mountains and joined the navy to get an education for a better life. The Argentine navy sent our cook to navy cooking school where he learned to prepare 10 evening meals primarily for officers. He was good at his job, so we voted to give him Sundays without duties and we alternated turns as the Sunday cook.

Dining room meal time.

After our noon meal, we discussed the work that needed to be undertaken that day. If the weather was good (that is, no wind, snow, or fog), we might take one of the small boats and visit different locations for a dive. Our dives were often brief, only 20 to 30 minutes, due to the extremely low temperatures of the water, usually minus 1° to minus 2° Celsius. During each dive, we collected different species of algae and recorded the depths for each. Specimens were collected in net bags and transported back to the lab.

The algae were sorted, recorded, studied and pressed between blotters to dry in a botanical plant press. It took 3 to 4 days to process each collection. The algae were studied microscopically in the late afternoon and we shared new discoveries.

The divers could take a shower if it was their turn. Showers were allowed once a week. If you wanted to wash your clothes you would take them into the shower with you. Of course, the "Guardia de Agua" had to anticipate some showers so he had to keep the large water tank in the attic filled with shoveled snowmelt water. After each dive, Waterhouse and I cleaned the wet suits and hung them to dry. If we needed to use up some energy, we would help the "Guardia de Agua" pump some melted snow water up to the ceiling tank.

Water

One of the daily tasks of which the scientific party all took turns was "Guardia de Agua," or water guard. The regular Argentine crew was exempt from this task. This duty entailed making sure that an adequate amount of water was pumped upstairs into a holding tank to last for an entire 24-hour day. The process was simple. We shoveled snow from outside the base building into a box 2'x2'x6' mounted on a ski sled. When the box was filled to capacity we could slide the

Me shoveling snow into the sled to bring to the caldera.

snow-filled sled close to an outside caldera that melted the snow, which was then pumped from the caldera to the holding tank in the attic.

Earlier we had measured the holding tank and determined that it could hold about 300 gallons and at about 8 pounds per gallon, which came to over a ton of weight. We calculated that one filling of the holding tank would provide enough water for the six scientists and the requirements of the cook to cook and clean the kitchen. The total consumption of water was 50 gallons, per person, per day. This is two times the amount of water we were allowed on the icebreakers.

I need to include here another rule regarding water use. The scientific crew had previously agreed that each member would be allowed one shower or one clothes washing each week. If we were diving, we could have a shower after diving. This rule was established to conserve water.

We also invented a dipstick to accurately determine the depth of water and thus its volume at any time by crawling up into the attic, with a flashlight and dipping the stick. Additional daily water usage for the lab use, showers, washing up, and laundry quickly lowered the water level in the holding tank. In fact, on several occasions the tank became empty.

Water consumption had to be regulated. Otherwise if two or three divers took a shower after a dive and washed their diving suits and regulators, this would quickly empty the water tank. The Guardia de Agua had to get more snow from around the base, fill the sled, shovel it into the caldera, and pump it to the holding tank in the attic. This effort was incredible and consisted of a substantial workout throughout the day.

One day we discovered soot floating on the top of the water in the caldera and the holding tank. The water was coming via the taps in the bathroom and in the kitchen. So, the entire contents of the holding tank, about 150 gallons, had to be drained leaving an empty holding tank for us to clean.

Once clean the process started all over again. We discovered that small droplets of particulate matter, like tar, coming out of the chimney had landed on the snow and absorbed radiation from the sun and melted tiny vertical tunnels in the snow. The particles would stop at some point and reside in the snow in various amounts. When we shoveled snow to fill the box sled the melted snow, filled with these black particulates, which floated, was pumped to the tank in the attic.

This entire project took several days to fix. The snow collected and shoveled into the sled closest to the base building had the most tar balls. By moving 30 or 40 yards away from the base there were no more tar balls in the snow. We figured out that a very rich diesel fuel mix for stoves, heaters and generators caused some of the fuel to be unburned and it was carried out the chimney with the smoke. The fuel supply to the furnace was adjusted and the problem solved. However, to obtain clean snow we had to continuously move further and further away from the base. Our advantage to this particular job was that you could take an extra shower.

The Washing Machine

The main building on Melchior Base provided accommodations for the scientific party of seven people including four Americans, two Argentines, and one Frenchman. Each had their own room roughly 6 feet by 12 feet with a bunk, wall locker, fold-up desk, chair, and a small bookshelf above the desk. About 72 square feet of space was plenty for each person. The rest of the building had seven bedrooms for personnel, two bedrooms for lab space, a dining/living room, kitchen, storage room, a small bathroom with toilet, sink and shower, and a small area with a navy surplus (grey) washing machine fully equipped with dual rollers to squeeze excess water out of the clothes. A small, deep sink was nearby to rinse clothes after they had been washed by the twisting and turning caused by an old fashioned three-blade turbo (this type still exists today).

The scientific group that normally resided in the main house negotiated the rules for the use of the washing machine. The rules were simple: if your clothes were dirty, use the washing machine and make sure it was cleaned up and unplugged when you were finished. The machine was electrically powered.

We had no specific time limits for a clothes washing, but usually we allocated 30 to 40 minutes of washing, then squeezing in the rollers, then soaking in the sink filled with clean water, then squeezing again, finally hanging the clothes on some overhead clothesline to dry. It took almost a day (18 to 20 hours) to dry. Ironing was not an option or needed.

The procedure for washing was to fill the washing machine enough to cover a reasonable pile of dirty clothes with warm water, add some powdered soap, the more the better, plug in the machine, and start up by the shift lever on the side. Then we would watch the operation for a few minutes and let the washing proceed. Not exactly a big deal. But, all of us appreciated having this modern convenience of a surplus WW II washing machine available. What could possibly go wrong?

Well, just imagine seven men with tons of dirty clothes, shirts, underwear, long johns, socks, and sweaters working around divers and slimy fish and bird poop every day—we had lots of dirty clothes. Anyway, you get the picture. If someone chose to do a washing, other crewmembers could add an item or two of clothing without being noticed.

It became routine if René was washing shirts, pants, socks and underwear, his clothes really filled the machine. Several people sneaked additional items into René's washing. Mostly the add-ons included underwear and socks, but sometimes a shirt or sweater was slipped in.

The machine, chock full of his clothes, was started early one morning. As soon as René left the washroom, someone else added two pair of waffle weave long johns. Other members of the crew added more laundry. Shortly thereafter, we all smelled an acrid electrical odor coming from the washroom. The excessive amount of clothing had become an entangled knot, the machine stopped, the turbo was immobilized and smoke was pouring out of the motor on the lower part of the machine. We removed the clothes and René quickly observed that many additional pieces of clothing were not his. The mechanics removed the motor, took it apart and pronounced it dead. We were all grumpy that night. By morning all the extra additions of clothing were gone.

We were now faced with the problem of how we could clean our clothing. The solution quickly became obvious. We could throw our dirty clothes on the floor during our weekly shower. If we were diving there was a slight advantage. Another option was to put our clothes in a plastic mesh bag and have it washed when we travelled around by ship. That was not very often.

By the end of the season at Melchior, all of the clothing of the scientific and naval crews had become disgusting. A huge bonfire took care of all our dirty clothes, parkas, boots, and underwear. We left with only the clothes on our backs.

Human Adaptions

Every individual is adapted to a biological clock and conducts their activities according to that clock. Typically, we think of working during the daytime and rest/sleep during the night. This type of cycle is called "Circadian Rhythm." Our team quickly learned that after dinner, usually 8-9 PM in the Argentine world, we still had plenty of daylight. With dinner over at about 10 PM, we often worked in the laboratory or engaged in conversation and music in the dining area. The music was mostly classical guitar played by the base commander, Lt. Cueli. I joined in sometimes with Woody Guthrie and Pete Seager folk tunes. After such a session, our clocks often said it was midnight or 1 in the morning, but it was still daylight outside.

Circadian rhythm is a daily rhythm based on a 24-hour clock. All organisms in Polar regions are adapted to a change in the day-night regime. We expect night time to be dark and daytime to be light. When it is dark we sleep, when it is light we work, much like a traditional farmer. With the invention of the incandescent light bulb, modern society has been able to extend the light portion of a 24-hour day.

For example, summer baseball can be played during normal daylight. It can also be played at night due to lighting technology which can light up a baseball park as bright as daylight. As I said before, the extension of summer daylight in the arctic means that some vegetables like cabbage, lettuce, tomatoes, and radishes can adapt to increased daylight and produce oversized examples of each vegetable. This works because green plants can combine CO_2 and water in the presence of daylight and the green pigment chlorophyll A, into glucose ($C_6H_{12}O_6$). The glucose, a sugar, serves as the energy source for photosynthesis, which supports the growth of the plant cells.

In Antarctica, six months of the year are in darkness and six months of the year have daylight. The growth of plants has to be complete in a half year. This means that the actual growth rate is faster when daylight is abundant. Reproduction of new individuals also takes place before the darkness arrives. Because photosynthesis takes place only in the presence of light, an abundant supply of glucose can be stored by algae which respire only minimally during the darkness.

Although we were not vegetables, our personal habits, such as sleeping or being awake, changed with light day and night.

As time passed quickly, we were now at mid-February, we noticed that our individual 'Circadian Rhythms' had changed. Since there was no night time darkness, we found ourselves eating dinner later and later in the day. Often, we sat down to eat around 9 or 10 in the evening. We did most of our diving and collecting in the late afternoon. When we returned to the base after a dive, Lamb, Zimmermann, and Delépine quickly took specimens of algae to study under the microscope in the main lab. The algal specimens were pressed up on herbarium sheets at this time. The custom of afternoon tea or matté gathered the scientific party for a little break.

Back at work, my job was to spread all the neoprene diving suits and other diving equipment out to dry. The diving tanks were filled for their next use. Since I had become quite interested in the taxonomy and ecology of the algae, I collected and preserved my own collection for future reference.

Our study of the algae usually took place from 4 to 9 pm. When the work was completed, we cleaned ourselves up, took one of our weekly showers, and dressed comfortably for dinner. After dinner we conducted lab work, read, engaged in radio communication, fixed something, or played cards and talked. If the conversation was enjoyable we often discovered that we had stayed up until 2 or 3 AM. Sometimes Martin and I would develop film and print postcards for the crew. Cueli was a classical guitarist and I was learning folk songs on the guitar I had purchased in Buenos Aires. Anyway, all of us were sleeping later and later than usual, sometimes as late as noon. Therefore, breakfast became lunch and lab work followed and more diving in the late afternoon if needed.

In any case, since there was no night time, our individual Circadian clocks simply seemed to move forward by several minutes each day. Some of us put black plastic over our bedroom windows which helped to get to sleep quickly.

Chapter 4
The Algae

"Every time you dive, you'll see something new—some new species. Sometimes the ocean gives you a gift, sometimes it doesn't."
James Cameron

OUR FIRST DIVE IN ANTARCTICA was at Melchior in the small caleta adjacent to the emergency house. We kept our diving suits, SCUBA tanks, regulators, and compressor there. Entry into the water was via a stone stairway, from the emergency house to a platform about 4 feet by 4 feet at the water's edge. It was easy to slip into the ocean from this

Here I am collecting algae.

platform. Our job was to swim around the entire harbor using just a snorkel. The point was to observe what the bottom was like and get a general view of the algal populations. It was an easy swim and our ¼ inch neoprene diving suits kept us quite warm.

I was a little cautious about snorkeling in a place that few people had ever seen, much less dived in. I kept a close eye for sharks, eels, and seals. I hardly noticed the images that were zooming back and forth around me. But I did eventually see that these images were penguins swimming by and checking me out, mindful that I could have been an aggressive seal looking for a meal. The penguins themselves were not threatening at all and I quickly accepted the fact that they had let me into their space. Apparently, I offered no threat to them.

The next thing I observed was that the bottom of the harbor was devoid of any large algae. Instead, the bottom showed a distinctly calcareous pink, crusted covering that looked a lot like *Lithothamnium* and *Lithophyllum*, quite similar to species I was familiar with from our algal studies in Rockport, MA. A foliose specimen of *Leptosomia*, which did have a tissue structure, was growing out of a crack in the rocky bottom. At first glance it appeared that all the fleshy algae had been scraped from the shallow depths down about 10 feet below the surface. I also observed several limpets, or single shell mollusks, which at first I thought had grazed off all the fleshy tissue from the algae from the upper rocky slope. Not so. I learned later that week that when small icebergs blew into the caleta from the northeast, the ice simply caused abrasion action to the rocky bottom and removed most of the algae. *Nacella polaris*, the limpet, cleaned up the algae that remained by scraping the plants with an abrasive tongue like structure called a radula.

Marine algae are unique plant-like organisms that have a special pigment, chlorophyll A, that enables them to absorb radiant energy from the sun and convert that energy into a type of sugar, called glucose. The algae lack a vascular system (no xylem or phloem) and are supported by the water in which they grow. The algae are separated into groups according to pigment array and structural components. The size of algae ranges from single cells (with or without flagella) to large kelps. Algae can be separated into four major groups.

- Cyanobacteria—Microscopic, bacteria like cells with photosynthetic pigments. They grow as coverings on rocks in shallow water and intertidal areas. They were found in areas, which were heavily grazed and scoured by ice.
- Chlorophyta—Small green algae with chlorophyll A. They live mostly in shallow areas, which have often been scraped away by the movement of icebergs and mollusk grazing.

- Phaeophyta—Brown algae containing chlorophyll A and C. These are strap-like kelp and large (sometimes 10 to 12 feet in length). They live at a depth of 10 to 40 feet.
- Rhodophyta—Red algae with chlorophyll A and phycoerythrin, a red pigment. These species are adapted to deeper water and tend to be less than 12 inches in height. Some red algae form pink crust. They are from the intertidal to deeper water about 80 to 90 feet deep.

The distribution of these pigments regulates the ability of the algae to utilize the sun to produce energy for growth and reproduction. The diversity of algal species is dependent on depth, sunlight, and temperature. Populations of algae have their greatest diversity in the tropics, with numerous species. Diversity declines toward polar areas, such as the Arctic and Antarctic. Little is known about the diversity and distribution in Antarctica. The severe environment, harsh temperatures, dark winters, and geographic inaccessibility have allowed only minimum opportunity for serious study. Our team was one of the first organized groups to use SCUBA to observe, collect, identify, and evaluate *in situ* algae in Antarctica.

In addition to the distinction of algal groups by color of their individuals, it is valuable to learn about their specific habitat characteristics. Studies have looked at which species live at which depths and when and how do these species reproduce and distribute offspring. Our job was to learn as much as possible about the Antarctic algae.

We had an excellent knowledge of the diversity and ecology of algal species from the study we had completed in Rockport, Massachusetts. With that background, we were able to evaluate the similarities and differences of the algal populations between both locations. One of the differences that was quickly obvious was the lack of species in the shallow areas and intertidal zone. The impact of small icebergs scouring the intertidal and shallow zones was dramatic. The abundance of vegetation in New England intertidal zones was impressive when compared to Melchior, where algae were completely absent.

In both locations, from 15 to 25 feet deep, brown algae, including kelp-like species, were most common. Calcareous algae occupied much of the bottom in shallow areas around the Antarctic Peninsula, but less so in New England. Deeper zones for both areas revealed an abundance of red algae.

At a depth beyond 10 feet, the algal vegetation was very rich by both diversity and size. Most conspicuous was the large, kelp *Phyllogigas*, which was similar in external appearance to kelp found at a similar depth in northern New England. *Iridaea* was a large, sheet-like red alga similar to the New England "dulse" or Rhodymenia. There were many other species of red and brown algae on the bottom, which I could see at a depth of some 30 feet. I was able to make a quick evaluation that the bottom was rich in numerous species of algae and suitable for our preliminary SCUBA dives and collections. My first time in the water lasted about 30 minutes, the water temperature was 1°C and I was starting to feel the cold. After noting the conditions here I looked forward to my next dive using SCUBA tanks.

This is a photo of large kelp.

This is calcareous algae on a rock.

We gather kelp at Deception Island.

Here we photograph the algae before we press it.

Though we did not quantify invertebrate animals our observations showed a surprising number of species. By the end of our time at Melchior we had completed nearly 40 dives using SCUBA. We visited several other stations and islands, including Almirante Brown, Palmer, Port Lockroy, where Lamb was stationed for over a year during World War II, Deception, and the South Shetland Islands. Some 75 different species of marine algae were collected, identified, pressed, and preserved with ecological data for each species.

Some of the species we observed and collected included the following. Many species names are presently being studied and thus not yet clear.

Chlorophyta—green algae

- Monostroma hariotii
- Enteromorpha bulbosa
- Urospora penicilliformis
- Urospora mirabilis
- Lambia Antarctica
- Monostroma applanatum
- Antarctosaccion applanatum

Phaeophyta—brown algae

- Ectocarpus
- Geminocarpus
- Phaerus
- Adenocystis
- Phyllogigas
- Cystosphaera

Rhodophyta—red algae

- Porphyra
- Delisea
- Lithophyllum
- Lithothamnion
- Kallymenia
- Curdiea
- Plocamium

- Gymnogongrus
- Gigartina
- Iridaea
- Leptosarca
- Antithamnion
- Ballia
- Georgiella
- Delesseria
- Myriogramme
- Pantoneura
- Antarctocolax
- Polysiphonia
- Ptilonia

The Botanical Survey of the West Antarctic Peninsula was completed successfully and safely with no injuries to the divers. Thirty-one genera and 34 species of benthic marine algae occurring on the coast of the Antarctic Peninsula and adjacent islands were described and illustrated. The algae were collected by the use of self-contained diving techniques (SCUBA) during the austral summer season of 1964-1965, the base of operations being the Melchior Islands off the west coast of the Antarctic Peninsula (64°19′S and 62°57′W) where all facilities for the expedition were made available by the Argentine Hydrographic Service.

Chapter 5
The Animals

"Antarctica. You know, that giant continent at the bottom of the earth that's ruled by penguins and seals."
C.B. Cook

BEFORE WE LEFT CAMBRIDGE for Antarctica Dr. Lamb put together a simplified food chain to show, in jest, how serious our daily encounters with local wildlife could be. The bottom of the food chain is occupied by marine phytoplankton at the microscopic level. These organisms are so small that a microscope is required to see individuals. Most of the phytoplankton are simple cells, filaments and occasionally branched filaments consisting of just a few cells attached together. Also, among the phytoplankton are tiny silica encased, photosynthetic unicells called diatoms.

Although there are several taxonomic groupings of phytoplankton, they are related by the fact that all of them have the pigment Chlorophyll A, which is used to form a simple sugar compound, Glucose, which is a source of energy for single-celled, non-photosynthetic animals known as zooplankton. In some cases, the zooplankton can move around in the water column to seek out the glucose rich phytoplankton.

The phytoplankton also produces oxygen and may account for as much as 70% of the oxygen animals on the planet earth breathe. Both phytoplankton and zooplankton are consumed by Krill, small crustaceans such as *Euphausia superba*, a species of shrimp found in the ocean around Antarctica. The Krill are consumed mostly by bottom dwelling fish such as *Notothenia neglecta* and *Chaenocephalus*. Fish in general could be consumed by SCUBA divers, "god forbid," because this is a terrible, bony fish and ugly as hell.

We were really worried when we analyzed the next link upward on Dr. Lamb's theoretical food chain. It shows a Leopard seal, 1500 pounds or so, with its giant teeth. Now we knew we were safe because there were no reports in the literature indicating Leopard seals have a preference for SCUBA divers. Their diet consists almost completely of penguins they encounter on the edge of the pack ice or near rocky outcroppings. But there are reports of orcas capturing and eating both Leopard seals and the much smaller, 150 pound, Weddell seals.

Smaller organisms in Antarctica include, of course, bacteria. These organisms, though microscopic, require a specialist to evaluate. We can say that bacteria have three basic forms coccoid (round), bacillus (elongated), and spirillum (spiral) shapes. From the three bacterial structures a group called the Cyanobacteria have evolved. These cells are among the first cells to exhibit photosynthesis. In photosynthesis, atmospheric CO_2 plus H_2O combine in the presence of Chlorophyll A and a few other pigments and RES or radiant energy from the sun, to produce glucose, which serves as a food source for the zooplankton, and the oxygen evolves into both the water and into the atmosphere, thus the start of the food chain.

The phytoplankton and zooplankton species are consumed by microscopic animals, including isopods, amphipods—krill/shrimp which are strained by the baleen plates of some species of whales. Humpback and Fin whales exemplify this type of feeding. Orcas are toothed whales and feed by biting down and crushing their prey of seals and penguins, then engulfing them. In the middle of the food chain there are large seals with large, pointed teeth that eat penguins and fish. These are the leopard seals.

It is interesting to note the pathway of single-celled organisms: eaten by krill, eaten by fish, then eaten by penguins, seals, and eventually whales. Whales are among the largest mammals on the planet.

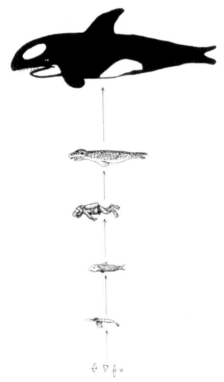

Lamb's simplified food chain

Penguins

The signature animal of Antarctica is the penguin. Their unusual shape and behavioral features separate them clearly from all other birds. On one occasion, Mack walked away from the base toward the northern point of the island. He set his backpack down on the rocks so he could take some photographs. The drawstring on the top of the pack was drawn closed but not tied securely. Mack moved several meters away and was crouched down looking at some geological formations at the shoreline. Several chinstrap penguins observed the knapsack and hopped close to get a better look (all penguins are nosy). The lead penguin unraveled the knapsack drawstring and pulled out Mack's special, double gradient density sunglasses. The penguin tried to move away with the temple of the sunglasses in his beak. Indeed, a funny looking penguin posing as a Hollywood movie star fiddling with the sunglasses. Mack's eye caught the action and he quickly returned to retrieve his expensive glasses.

There are about 17-18 species of penguins that exist in the world. Their collective geographic range in the Southern Hemisphere is from Antarctica to Argentina and Southern Chile. Some species even extend their range to include parts of Australia, New Zealand and southern Africa. The Galapagos Islands have a rare species, *Spheniscus mendiculus* that lives and breeds on the equator. This is most likely due to the cold Cromwell Current, which provides cold-water nutrient-rich food.

There are three species of penguins that we saw most frequently: *Pygoscelis antarcticus*, the chinstrap; *P. Adelie* the Adelie penguin; and *P. Papua* the Gentoo penguin.

90

These penguins have a lot in common and share numerous structural characteristics, but they do not interbreed. They are around 18 inches tall and weigh about 10 pounds. They, like other species of birds, have external contour feathers and tiny, fluffy down feathers closer to the skin. The contour feathers offer some protection from the weather and they provide insulation from the cold. They have actually formed "de facto" down parkas with the down inside the contour feathers. All three species have a formidable beak, which is adapted to remove salt from seawater, and to crush keratin encrusted krill, *Euphausia*, their primary food. Their wings are smaller compared to other birds and are flightless yet adapted for swimming. There are estimates that these three species are able to swim at speeds of 25-30 miles per hour.

At the base of their spine some feathers have evolved, becoming thickened and serving as a tripod-like support for extended standing. Their bones are solid and heavy which contributes to their diving skills. The penguin's feet are very fat and calloused on the bottom. The fat serves as insulation against the cold rocks of their nests. The feet have three digits and they are webbed for swimming. Most have a smaller digit on the inner side of their feet. Between swim wings and webbed feet, they are adapted for escaping their prime predator the Leopard seal, by rapid swimming and being able to jump quickly out of the water.

One other point to mention is the penguin's ability to regurgitate on demand. The reason for this skill is for the penguin to feed its young. When preparing to supply food to offspring penguins jump into the water and swim around rapidly looking for shrimp. Then the shrimp (krill) are gobbled up and partially digested. The penguin swims back to its mate that is guarding the chick. The chicks turn their heads upward and open their mouths wide. The penguin with the food coughs it up and deposits it into the open mouths of the baby chicks. Typically, there are only one or two chicks.

The penguins can regurgitate on demand and can also use this feature to ward off threats. When threatened by humans, they produce projectile vomit which covers any potential predator. They also have no cloaca to close their digestive tract so wet fecal material often drips from them when they move. (Pigeons and seagulls also do this.) The smell of a penguin rookery is awful and the noises penguins produce is disgusting. The role of rapid recycle-ing food by penguins is important to the Antarctic nutrient cycles because proteins are extracted from Krill, digested, and then waste re-enters to provide nutrients to the phytoplankton.

Penguins on the edge of the pack ice and on the rocky shore are able to move about by rapidly jumping from place to place or they walk around with a side-to-side waddle. When penguins move from sea to land they swim rapidly and actually pop out of the water, sometimes 5 or 6 feet high, always landing on their feet.

Some researchers have studied the issue of mate selection and fidelity among these penguins. Males tend to "pick up" lone females and present the female with a small stone (like an engagement ring). This can take several days of looking and presenting a stone. If the female accepts the stone, the ceremony is over and the honeymoon begins. Once the honeymoon is over both penguins are faithful to their partners for life. In fact, the divorce rate is lower among penguins than humans. Eggs are produced (1 or 2) and the female incubates them for about 34 to 36 days. When they are hatched, the new penguins are very ugly. They are balls of dirty, gray fur, always hungry and constantly crying for food. Parents take turns feeding and caring for the young for about 50 days, then teach them to swim and catch food. Then they are on their own. In 3 to 7 years they are ready to look for a mate and start breeding.

Chinstrap Penguins at play.

The Chinstrap

Typically, the Chinstraps are about 18-20 inches tall. They, like other species of birds, have long but small external contour feathers and tiny, fluffy, down feathers closer to their skin. The contour feathers offer some protection from the weather and they enclose the small downy feathers, which provide insulation from the cold weather. The Chinstraps also have a formidable beak, which is adapted to remove salt from seawater. The beak is able to crush krill (*Euphausia superba*) and digest them. Their wings are proportionally smaller compared to other birds and are adapted for swimming. Even though they try they cannot fly.

At the base of their spine some feathers have become thickened and serve as a tripod-like support for standing. Their feet are also quite large, webbed and show a lot of fatty tissue. The foot tissue is, I believe, functioning as insulation. Swimming speed is from 25-30 miles per hour. Unlike other birds their bones are solid and heavy, an advantage for diving at which they are adept.

The Adelie

The Adelie penguin species is similar to the Chinstrap penguin in nearly all recognizable characteristics. They are, however, distinguishable from other species by their almost complete black coloring of their feathers except for their stomach areas and the underside of their flippers, which are white. Like Chinstraps, their tail feathers are rigid serving as a tripod and providing extra stability. The feet are calloused, fat, webbed, and have fingernail like claws at the end of their toes. I also noted that there were far fewer Adelies compared to Chinstraps around our base.

The Gentoo

The Gentoo penguin species is present in smaller numbers than the Chinstraps and the Adelies. All three species are about the same size and seem to differ primarily in their external coloration. Gentoos have some red coloration on their beak and white patches above their eyes. They are easily distinguishable from Chinstraps and Adelies. They do, however, also have rigid tail feathers to give them the stability advantage of a tripod.

Penguin species were limited to the three described above. Their size, weight and general habits are similar. I am surprised that no evidence of inbreeding has been suggested. The DNA for these species should be evaluated if not already underway.

All penguins are cute and are often a major attraction in aquariums. Their colors are restricted to tuxedo like black and white. We saw penguins every time we were diving. They are nosy and smell bad. If you sat down on a rock amongst them and do not move rapidly they would come very close, but remain guarded. We learned that they liked roasted peanuts.

In his journal, Lamb recorded an interesting encounter with a penguin:

Dec 1, 1964

We do not appear to have an alarm clock in our store, but in the case of an expedition member who was sleeping rather late, the problem was resolved by catching a chinstrap penguin on the small ice capped peninsula off the end of the station promontory and introducing it into the sleeper's cabin.

My account was slightly different:

Dec 1, 1964

Today I slept late and awoke to find a penguin in my room. This was a subtle reminder to wake up earlier. The penguin went nuts and all hell broke loose. The penguin quickly realized that my room was too small for both of us. I opened my door and left.

Seals

Swimming and SCUBA diving when surrounded by seals was somewhat strange when our group first entered the water. We were all trained to be cautious of underwater critters like conger eels, spiny sea urchins, lobsters, and sharks. None of us cared to be bitten, stung, or punctured. But the Weddell and Leopard seals were completely new swimming partners. The Leopard seals looked evil to me because of their big eyes and exceptionally large teeth. To me those teeth looked like those of a dog that bit me when I was a small kid. The Weddell seals also had substantial teeth but they seemed to project puppy-like grins as they swam amongst our dive team. We never feared being bitten, but sometimes the unexpected site of a Leopard seal raised the hackles on the back of my neck. Our divers never had a frightening encounter with any seals, but we did maintain a level of vigilance at all times.

Seals are commonplace in Antarctica. We saw them swimming and playing along the shoreline of our base every day. Seals are completely adapted to the marine environment. Their body shape is fusiform and well designed for smooth passage through the water. Seals have thick layers of fat, which serves to insulate their bodies and allow them to retain a good portion of their body heat. Fur also compliments their fat and increases their ability to retain heat.

There are about six species of seals that live in various parts of Antarctica. Of the six species, only two were common around our base, the Weddell seal, *Leptonychotes weddelli,* and the Leopard seal, *Hydrurga leptonyx.*

The Weddell seal is the most popular, curious and playful. They range from 150 to 200 pounds and are about the size of a man. They swim in groups and seem to enjoy the company of SCUBA divers. They also like to lay on small icebergs and sleep, especially after they have had a good meal of krill or sometimes fish. Krill, are the shrimp that breed abundantly in Antarctic waters. Weddell seals often followed our divers to satisfy their curiosity of collecting algae from the coastline. On more than one occasion seals followed us underwater and nibbled at the ends of our flippers as we were swimming up to board the dive boat. We could not decide if this behavior was just playful or flirtatious.

This is a cute Weddell Seal. Photo by M. H. Zimmerman.

The second most common seal that we saw on a daily basis was the Leopard seal. The leopard seal was not playful or flirtatious, but just the opposite—plain mean. Instead of feasting on krill, the leopard seal was a top-of-the-food chain carnivore and predator. Their size was large, in the range of 1000 or more pounds and about 10 to 12 feet in length. Leopard seals have a fusiform body with a serpentine head and very large teeth. These teeth were well adapted for ripping, tearing, and killing. Their main prey was penguins, especially Chinstrap and Adelie. If necessary, the leopard seal could eat krill but that was not necessary due to the abundance of penguins around our base.

This inside out penguin was thrashed to and fro by a leopard seal until muscles were exposed and could be eaten.

It is worthy to note one leopard seal behavior that we saw frequently. Penguins gathered at the edges of the shoreline ice. Hundreds of them stood around on the ice or rocks looking for an opportune time to dive into the ocean in search of krill. The penguins were always on the lookout for the leopard seal predators. The heads of the penguins would turn to and fro, up and down, waiting for a time when no leopard seals were in sight. Then, penguins at the back of the pack began bumping their stomachs against those closest to the edge of the ice. Eventually the penguins on the edge would get pushed into the water. If the one or two penguins were not attacked and eaten by a leopard seal, the rest of the pack quickly jumped into the water and continued their search for krill. Penguins can swim much faster than seals and thus are mostly able to avoid becoming a seal meal. On the other hand, if leopard seals were present and managed to capture a penguin, the seal would bite the head of the penguin and swing it back and forth until the skin and feathers actually turned inside out and exposed the meaty, fatty tissue. Leopard seals are large, ferocious, and mean. They are also scary to divers and not playful like the Weddell.

This is a Leopard Seal skeleton. Note the large, sharp teeth.

Whales

The base at Melchior during our expedition was not the most exciting place in the world. However, there were abundant levels of excitement surrounding us. In addition to numerous penguins and seals, we saw many whales. They seemed to cruise almost daily past our base, we guessed, looking for food. For the most part, food for the whales was marine amphipods called Krill, or seals and penguins. Baleen whales prefer krill, while toothed whales prefer penguins and seals.

There were several whale species which could we could easily identify. Whales are separated into two groups based on different structures for feeding. *Odontoceti* are the whales with teeth such as the Orca (killer whales) and the Sperm whale. The teeth allow the whale to bite their prey—seals and penguins. The teeth are not used to chew, just tear out the meat. Thus, the teeth tend to be pointed rather than flattened.

The second group is the *Mysticeti* or baleen whales. The feeding structures of baleen are rows of several hundred keratin plates, a protein attached to the upper jaw of the animal. This feature allows the whales to 'filter feed' on small organisms. Whales make a deep dive and come to the surface with their large mouths open to engulf food like *euphausid* shrimp or small fish and eels. When they engulf food, baleen whales close their mouths with the baleen plates overlapping the lower jaw. Seawater is expelled from the whale's mouth by pressure applied with their large tongue. The food is retained and gulped down and digested. Some whale species are reported to consume a ton or more of food each day.

The whale species that we saw the most were the Orcas (killer whales); they are toothed whales. They usually passed by our base, swimming close to the surface in pods of three or four. I was always surprised at how close to shore they swam hoping to encounter a sleepy seal for lunch. They were easily recognized by the immense size of their triangular dorsal fin and black and white coloring.

We also saw several other species including *Megaptera novaeangliae*, the humpback, easily recognized by its frequent tail flash when making a deep dive. The humpbacks can hold their breath for more than 10 minutes. The fin whale, *Balaenoptera physalus*, is recognized by its immense length and small dorsal fin. Fins are the second largest mammals ever to live on the planet. Blue whales are the largest mammals. The smaller size minke, *Balaenoptera acutorostrata*, about 20-30 feet long and have a pointed forehead. These three species are difficult to distinguish from one another unless you are close to them. One thing the *Mysticeti* have in common is their ability to filter feed.

102

This is an Orca looking for a seal or penguin lunch. (Photo by Jen Kennedy.)

Other species of whales such as the toothed sperm whale, *Physeter macrocephalus*, and the baleen blue whale, *Balaenoptera musculus*, (the largest mammal on the planet) were never seen around our base.

Fishes

There are only a few species of fishes that we saw in Antarctica. Fortunately for us, Dr. Norberto Bellisio, an Ichthyologist from the Museum of Natural History in Buenos Aires, Argentina joined our group from the beginning of the expedition.

Dr. Bellisio was an expert in his specific field of Antarctic Ichthyology. He has made numerous expeditions to the Antarctic Peninsula before joining us. He was an extremely dedicated scientist and a pleasure to have in our group. His work entailed 'long line' mostly bottom fishing at depths of 58 to over 100 meters. Each day Bellisio rowed out of the caleta and placed his weighted long line, with about 100 baited hooks, in the water. We were surprised to learn that he was using raw beef for bait. He left the baited long lines in the water for 24 to 48 hours then returned to retrieve his hooks. There were two species of fish that he commonly caught, *Notothenia neglecta*, the ugliest fish in Antarctica, maybe the ugliest fish in the entire world.

Drawing of Notothenia neglecta by K. G. Fralick.

There was another species, the genus *Chaenocephalus*. This was second in the line-up of the world's ugliest fish. Since they live near the bottom it is difficult for the fish to find food. It seems they may occasionally find dead krill such as *Euphausia*. So, the prime beef bait never failed to do the job hooking a fish. The fish have very large heads and a body that decreases in size from head to tail. They have large eyes and a gaping mouth. In addition to the krill we often found nemartine worms hooked up in the long line; they came up in an entangled mass as big as a basketball. The worms were white, slimy and long. I unraveled a few of these worms and found them to range from 15 to 20 feet long. I never saw any of this type of creature while diving in Antarctica.

Antarctic fish are the basis of numerous biological studies because they have a unique circulatory system. The blood of such fish appears to have an anti-freeze capability which prevents a "freeze up" when the surrounding water temperature is below freezing. At a temperature of -2°C, blood should freeze into ice crystals, but it does not freeze in these uniquely adapted fish.

Anyway, Dr. Bellisio collected his fish, returned them to his lab to count, weigh, and measure. He also dissected the liver or 'higado' of many fish and studied them. Bellisio was the brunt of many jokes because he preserved his specimens in Formalin, which wafted through our living quarters. Formalin is a watered-down form of Formaldehyde or embalming fluid, used as a preservative in laboratories. The odor of formalin is putrid and worse than a sniff of household bleach. It made my nostrils cringe.

On one occasion, as Bellisio was retrieving his longline, a leopard seal tried to get into the stern of the small rowboat that Bellisio used for fishing alone. The seal was really large and weighed an estimated 1500 pounds.

The seal finally moved away from the boat when Bellisio conked it over the head with the oar.

Birds

During our time in Antarctica we were fortunate to have exposure to numerous species of birds. Of course, we were able to anticipate the penguins in advance of our visit, and were delighted when we first saw them. No birds can compare to the personalized moves of penguins unless you sat still and observed their behavior for several hours. But, there were other birds that attracted our attention every day.

Paloma Antarctica soaring over the ocean.

The first on the list of common Antarctic birds was the Albatross, first seen off the stern of the *San Martin* shortly after we left Ushuaia heading into the Drake Passage. An

Albatross followed our ship, flying behind for 600 miles over a three-day period. We never saw that bird, with a 10-foot wingspan, stop flying for any reason—seemingly with no food and no sleep, only remaining on course.

Once we were on land from Deception to Melchior and further south, we were never beyond the reach of cormorants, terns, and skuas. When we were diving, we saw Wilson's petrels dangling their feet just above the water. On occasion, we saw a few white doves, una *Paloma Antarctica*. Several indeterminable species arrived when any garbage or food waste was thrown overboard into the sea.

Chapter 6
Dangerous Situations

"Difficulties are just things to overcome, after all."
Ernest Shackleton

THERE WERE SEVERAL DIFFERENT TYPES of dangerous situations that we encountered over the course of our trip: weather, animals, mechanical failure, white-outs, and getting lost. The last potential danger was diving in unfamiliar territory and getting lost on our small boat excursions far away from contact with our base. No GPS existed back then and our radios were unreliable.

Dangerous Weather

The first part of our expedition involved crossing the infamous Drake Passage, a distance of about 600 miles, from southern Argentina (Ushuaia) to the Antarctic Peninsula. We already knew that an icebreaker ship was specifically designed to penetrate through the coastal pack ice such as around the Antarctic Peninsula. The ship's design incorporated a blunt rather than a pointed bow, which worked efficiently in the ice. I was concerned that the ship had no keel to speak of; this made the icebreaker less seaworthy in the rough ocean conditions.

After a few days in Ushuaia we set out for the 600-mile crossing of the Drake Passage to the Antarctic Peninsula. Before leaving Ushuaia, the ship's fuel and water tanks were filled and all deck cargo was carefully secured. We had only traveled six or seven hours when the ship slowed down and dropped anchor behind Picton Island.

We became quickly aware that the weather beyond Picton was extremely windy and the seas agitated so we remained at anchor there for two days. Finally, after what seemed like forever, we hoisted anchor and were underway.

The weather must be improving, we assumed. Nope! Once clear of all the islands, we were immediately exposed to the open sea in the Drake Passage. This area has been referred to by early New Bedford whaling captains as the worst ocean weather in the world. The ship designed to break ice was not at home in the rising seas that had swells of over 20 feet. The comments recorded by early whalers were clearly an understatement.

The ship's crew passed out heavy storm straps that we could use to keep us strapped in our bunks. Waves were breaking over the bow of the ship; they even crashed over the bridge. These waves were caused not only from the wind but also from the convergence of water currents coming together from several different directions. A person could not stand for more than a minute or so between waves. Crewmembers were getting sick and vomiting all over the ship. This was no surprise, since much of the crew came from the Argentine pampas or mountainous region and had never seen any ocean at all growing up.

As far as I was concerned I was a little nervous especially when I first saw 30 foot waves off our stern. The thought of getting swamped ran through my head. I checked to make sure my life jacket was stored under my bunk. Fortunately, I did not get sick to the point of vomiting. Fifty percent of the sailors got pretty sick and the companion way between my cabin and the bathroom was very slippery. All doors and portholes were ordered closed to keep the ocean out of the icebreaker. It was really strange to see the ocean cover our porthole as the ship rolled from side to side.

The cooks tried to serve meals in the Officers Mess. Here the chairs at the dinner table all had arms with chains to attach to the table. The table itself was designed with pop-up components into which you could insert a coffee cup, matté gourd and dinner plate. Even with this unique feature we could never keep food on the table. The evening meal turned into hardtack and wine.

By evening, the waves were about 25 feet in height. The best place to stay was in our own cabin strapped into our bunks. Occasionally we had to make a trip to the head (men's room), which was located about 30 feet from our cabin toward the stern of the ship. There were metal bars along both sides of the companionway to keep us on our feet. Bars were also in the bathroom to keep us somewhat steady, even so it was a slippery and smelly mess.

I returned to my cabin, strapped myself into my bunk and wondered if my life jacket was still under my bunk. Sleep was out of the question. Crunching up along the bulkhead side of the bed was the only option. I dozed off from time to time. One time I heard an unusual noise coming from the companionway outside my cabin. The noise was a loud clanging, something sliding and a lot of hissing. After a half an hour, I went out into the now dark hallway to find a large, heavy CO_2 fire extinguisher bouncing and sliding to and fro in the companionway. When the extinguisher hit the wall the handle also hit and released a blast of frozen CO_2.

Some sailors came around to investigate and eventually captured the flailing extinguisher. I went back to my bunk, but still couldn't sleep. Morning eventually arrived with bright sun, cool calm breezes, and 30-foot seas. Sometimes when the waves hit the ship sideways, we would roll up to 50 degrees from side to side. In fact, we could walk up the wall on one of these rolls. There were some footprints along the wall to show that these rough seas had been crossed before.

After several days at sea in the Drake Passage, with the winds and high seas, minimal sleep, and no real food, we were happy to anchor behind some islands at the top of the Antarctic Peninsula. The leeward side of the island gave the icebreaker and our crew a chance to rest, eat, shower, and get the ship cleaned up before we moved on toward Deception Island.

Another dangerous weather situation were the days of heavy fog, which often surrounded the ship and caused nearby coasts and islands to vanish before our eyes. The fog made navigating the ship difficult since most of the nautical charts were far from accurate. I should add that there are different charts for each section of the Antarctic Peninsula. These included Argentine, British, and Chilean charts. Names of islands and passages were different on each chart. This meant that each island or point of land could have three different names. The same chart overlap applied to all coves and charts of the peninsula.

Whiteouts

The weather eventually calmed down after two and a half days in the Drake Passage. The ship's course then took us into an ocean filled, extinct volcanic crater to stop and provide food, mail, and supplies to the Argentine base, Deception. The weather was damp and foggy. The entrance to the crater was narrow and shallow. The ship entered with ease. The ship anchored and we could see an active base with smoke emerging from the chimney.

There was no landing point other than at the edge of large slabs of pack ice about 6 feet thick. These slabs of pack ice moved around in the crater according to the direction of the wind.

The small cargo vessel from the *San Martin* had an engine failure after only one small load of cargo had been unloaded on the ice. While the engine was being repaired, the captains 'gig' was launched to bring scientists and crew exchanges, food and mail to the base. It was almost like a ferry from icebreaker to the base. When I took this ferry, I noticed the fog rolling into the crater where the ship was anchored. Within minutes of my landing ashore I could no longer see the ship. It was shrouded within a cloud of fog known as a whiteout. A whiteout occurs when ambient air temperature is warmer than the water creating fog, and when the fog blends in with surrounding snow and ice everything is absorbed by this white cloud fog.

The ferry was shut down for several hours before we could see the ship once again. It seemed like the ship had simply vanished and then reappeared three hours later.

Outboard Motor Conks Out

About 2 months into our research we were getting ready for another diving operation to collect algae. That morning I awoke early and filled the diving tanks. After lunch Lt. Cueli and I took the motorboat toward Alpha Island to check for leopard seals before Lamb and Zimmermann made the final dive.

The outboard motor conked out behind Tempano (iceberg) Katy Glennon. We rowed to the back side of Gamma Island where we quickly investigated a running fresh water source used by the early whalers. We also observed ropes, pipes, some rags and pieces of hose that were probably used to siphon melt water draining from the surrounding high cliffs.

Eventually, after many tries, we finally made radio contact with Melchoir base. People on the base sent Bellisio and Phillipe Aguilera after us in the remaining small boat. By this point the fog had become extremely thick. We saw Bellisio's boat and fired several rifle shots to gather his attention. After the third shot Bellisio finally turned and headed in our direction.

114

We were towed a mile or two back to Melchoir just in time to take Mack, Zimmermann, and Waterhouse for a final dive along the ecological transect we had set at the mouth of the Melchoir caleta. We could see the heavy fog off shore so we confined our dive close to shore.

Icebergs

Another potentially dangerous situation was the problem of floating icebergs. Often, we woke up in the morning to find the mouth of the caleta completely blocked by icebergs that had drifted in and became grounded. When this occurred, we were unable to take our little boat beyond the mouth of the harbor, nor was any boat able to enter. We were pretty much locked in. My biggest worry was the possibility of us being unable to re-enter if an iceberg blockage happened while we were afar and returned to find a pile of icebergs between our boat and the base. We also had a concern that our wooden boats could be crushed by some of the icebergs.

One large iceberg could also, easily come aground and block the entire mouth of the caleta. When large icebergs did arrive, we were usually bound to the base for several days at a time.

When we were able to bring our divers out of the harbor by boat we often saw seals sleeping on various icebergs. Sometimes these icebergs rolled over and dumped the sleeping seals into the water. We were always cautious of icebergs rolling over on us while diving.

Mechanical Problems

Our team had given only minimal attention to the possibility of mechanical problems prior to the early departure of our equipment in July of 1964. For example, we never anticipated even the possibility of a generator failure. We also did not expect to be limited to such small, twelve-foot long wooden boats. The Argentine navy had two outboard motors and they were not dependable. We also did not have GPS, as it hadn't been invented yet, or even an accurate compass nor up-to-date nautical charts. We lost several diving days due to failure of the outboard motors and lack of accurate coastal charts. Sometimes I am amazed we did not get lost or stuck in a whiteout or blocked by ice.

Leopard Seal

An example of an encounter with the ever-present Leopard seal occurred at the end of December. Early in the afternoon, as we were preparing for a dive and putting wets suits and air tanks on, we heard a commotion coming from the mouth of the caleta. Lamb described it as follows:

Dec. 30, 1964

As we were doing the (prep work), we heard some shouting and going out onto the steps of the emergency house, saw Dr. Bellisio on the shore rocks at the north side entrance of the inlet trying to fend off a huge sea-leopard which was swimming around his boat moored beside him, and now and then coming right to the water's edge where Dr. Bellisio was standing and rearing itself up out of the water as if it intended to haul out on shore and attack him. A very steep snow bank behind blocked his retreat and he was unable to get more than a meter back from the water's edge. He had no weapons and was shouting and pitching snowballs at the monster's head when it reared up towards him out of the water.

It was a huge animal, over 3 meters long, as long as Dr. Bellisio later recounted. From where we were standing, we could get a good view of its large, elongated, snake-like head, at least a half meter long, every time it surfaced. Lt. Cueli fetched a loaded rifle from the house and embarked in the other boat with Waterhouse and Gubolin to go to Dr. Bellisio's rescue. When they came within 5 meters, Cueli opened fire on the monster when it surfaced.

The first shot missed, but the second struck home and the beast quickly disappeared from view in a violent thrashing of water. A minute later it reappeared on the surface, about halfway up the station inlet to where we were standing. Cueli, Waterhouse, and Gubolin in one boat and Dr Bellisio in the other followed, but the animal failed to reappear again and they all landed at the emergency house jetty.

The dive was cancelled and efforts were made to locate the wounded seal. Later in the day, Cueli and others succeeded in killing the seal as it was lying on an iceberg at the entrance of the inlet, bleeding from a wound in the neck.

Lamb said in his journal, "Be this as it may, I would much hesitate to dive near where one of these animals was sighted, on the same day."

This is the Leopard Seal that attacked Dr. Bellisio in the Melchior Caleta.

Encounter with a Killer Whale

We arose one morning at 10 AM and had our usual breakfast of fresh rolls, cheese, butter and coffee. We also reviewed our day. The dive plan today was for Lamb and Waterhouse to make a deep dive to 135 feet. The bottom time was not to exceed 15 minutes. The planning and execution of such a deep dive made me nervous. A marked line had been set the day before.

Lamb, Waterhouse, and I were towed to the dive site by Cueli and Zimmermann in a second small boat. A large canvas collecting bag had also been placed at the bottom of the diving line. I set underwater watch bezels, and the divers grabbed a lead weight ring and mesh collecting bags and headed for the bottom just before 5 pm.

The divers found a dense covering of red algae and some large *Phyllogigas* plants, starfish, tunicates, like rugby footballs, and Crinoids with a peduncle and huge logghead sponges. A unique green alga, *Derbesia Antarctica* was also found. The depth of 41 meters 130 feet was reached. Samples of algae were collected and bagged. The divers stuffed the collecting bags into the large canvas bag. After slowly ascending the dive line they reached the decompression stop at 10 feet below the surface. After taking several breaths of air, the danger signal (four tugs on the line) signaled the divers to surface immediately.

I described it in my journal as follows:

"As the divers reached the ladder I grabbed the tanks on their backs and hauled them into the boat. The divers were one on top of the other tangled in a mass of arms, legs and diving equipment. I rowed madly toward shore."

Lamb's journal adds:

"The sea leopard must be following and trying to upset the boat. When we got close to land and the divers somewhat untangled."

Two orcas had probably, out of curiosity, came very close to us. So close, in fact, that I could have easily reached out and touched them. When the orcas blew air and my face got wet, I knew it was a close call.

While the divers were down at the decompression stop, Zimmermann suddenly heard the expulsive hissing blast of the two killer whales close by. The whales cruised over the marked line where Lamb and Waterhouse had been only two minutes earlier. Zimmermann said, "That was really a close call," his face very pale.

Zimmermann's journal gives by far the most detailed account:

"ORCA! GET THEM OUT QUICK! I have heard this sound before. The sequence of events that follows takes a mere 4 1/2 minutes, however it seems to us like a long time, and we shall certainly never forget it. It is absolutely clear to me that we have no chance to reach the shore in time. Killer whales are very much faster than our ridiculous little row boats. Furthermore, nothing would be easier for them than to tip over our boats; they have been observed to break 75 cm thick ice floes to get seals.

I turn our boat (I happen to be on oars today) in order to throw Dick the line. But he has already pulled the divers (who came to the surface with the four-tug emergency signal) into the boat. He first grabbed Mack, yanked him in, face down, then Richard on top of Mack. Before I have made the full turn and am ready to throw the line, Dick grabs the oars. I therefore turn again and we both row madly toward the shore. Cueli begs me to let him row, we change places and I keep an eye on the water surface. I expect the orcas any moment, maybe we'll all spill in a few seconds.

Cueli, the hardy Navy officer, is very pale. When we have moved some 50 meters toward the shore I see the orcas again, there are two of them. They are evidently not interested in us, they have made a sharp turn and travel now parallel to our shore from right to left. The divers' breathing and our rumbles with the oars on the boats, must have been conspicuous to them, they were so near. The scare still sits within our bones, and we continue to row toward a shallow cove.

Dick's boat, that had the divers, was ahead of my boat with Cueli. Four flippered feet can be seen sticking over the edge of the boat.

120

The poor divers have no idea what happened, they lie in the most uncomfortable position at the bottom of the boat on top of each other, face down. With every pull of the oars, Dick's feet dig into Mack's body.

When we reach the little cove inside Gallows Point we feel relatively safe again, we rest and relieve the divers. Now we can tell them what happened. Dick looks at his stop watch and makes the comment that exactly 4 and a half minutes have passed since we heard the breathing sound of the orcas. We row back to the station. Cueli, Dick and I are still in a state of shock, the divers, on the other hand, are rather cheerful, this annoys us. This psychological tension, which we begin to understand only gradually, remains with us for several days."

In Antarctica, the potential for dangerous situations is ever present. We could never be complacent with any situation nor were we fully cognizant of the potential for problems. All of us had to be ready to expect "glitches" every day and prepare in our own minds how we would cope with any adversity. All of us had read and admired Shackleton and knew that our problems with weather, animals, and navigation were negligible compared with his.

But, in retrospect, we did become a little overconfident toward the end of our expedition. For example, we did not wear life jackets every single time, nor did we always sign out for excursions outside our living quarters. Nonetheless, we were able to finish our fieldwork safely and efficiently.

Chapter 7
Other Bases

"The land looks like a fairytale."
Roald Amundsen

WE DIDN'T SPEND ALL OUR TIME on Melchior. When time, transportation, and opportunity allowed, we made every effort to visit other bases. One day the icebreaker ship *General San Martin* arrived at Melchior Island and gave us mail from home. A mechanic from the ship checked our extra generator and pronounced it dead (unfixable) so we would be required to operate Melchior base on only one generator running 24 hours a day with no backup. Zimmermann and Cueli set up the tiny generator, which Zimmermann brought with him to provide light for the microscopes.

The following day, the US Navy ship *Edisto* returned to Melchior, and Commander Nickerson offered our team a ride to Palmer Station (see map) and maybe even further south. The journey south to Palmer Station took only a few hours. *Edisto* set anchor as a small launch steamed out from shore to meet us. Palmer Station was formerly a British base called Port Arthur. The island was named after Nathaniel Palmer, a New Bedford whaler who discovered it in the mid 1800's. Once ashore a geologist and an entomologist (one who studies insects) gave us "the tour". The first place we visited was Skua Lodge.

Zimmermann at Skua Lodge.

The Skua Lodge was an old British Base on an island covered with ice, snow, moss plants, and some construction debris. There were also many species of Lichens present, as well as an Antarctic grass called *Deschampsia*, one of two vascular plants that lived in Antarctica. I suppose the supply of guano from the many skuas is what allowed such luxurious diversity of plants here providing them with rich nitrogen deposits. The skuas dove at Martin's head and scared him away.

In addition to the skuas there were also many screaming terns present. Terns migrated each season from New England and Southern Canada to the Antarctic Peninsula. This was their biological breeding ground. We continued to explore the base where a sturdy new building was being constructed by the US Navy Seabees. The fuel for the base was stored nearby in large, rubber pod tanks, which looked like they were bulging at the seams just waiting to be punctured.

After several hours ashore inspecting what would become the new American base, we returned to the *Edisto* for dinner.

Once back on *Edisto* we were able to explore the ship. They had a modern metal shop with lathes, drills, metal stands, and various metal work tools. After a great dinner of American chop suey, we were treated to two stupid, but enjoyable movies including popcorn. The movies were usually shown in the dining hall. Nearly the entire crew of the ship crowded into the dining hall, half facing towards a bed sheet hung up in the middle of the room. The other half of the crew was on the other side of the sheet. Therefore, half the crew was watching the movie backwards as the projector projected through the sheet.

I had to laugh because the sailors had learned to read many of the subtitles so the men on the other side could understand, as they couldn't read backwards. Of course, the sailors were very creative with procedure, especially with their rendition of text spoken by females.

Helicopter Ride

Another day, we rose for breakfast at 6AM, or 0600 in Navy talk. The ship was bumping and grinding in the pack ice. The navigator notified us of our position, which was 66:55 south by 67:05 west in the Matha Strait. Tall mountains were nearby and Lamb and Waterhouse prepared to visit some nearby mountain peaks by helicopter. A large Sikorsky helicopter called "The Horse" was wheeled out of its hanger on the ship. Lamb and Waterhouse, in their orange jumpsuits, boarded the chopper with knapsacks filled with collecting tools to chisel lichens off the rocky mountain peaks, and a supply of candy bars for emergencies.

The chopper looked like an Army surplus from WW II and was painted orange to make it easy to find if it crashed. The life expectancy if the chopper fell in the water was four minutes.

When the large helicopter returned to the ship after about an hour, Martin and I donned the orange survival suits and climbed into the smaller Bell helicopter. We took off and circled around the ship giving us a fantastic aerial view of the ship surrounded by broken up pieces of the ice pack.

Helicopters are amazing, especially the small ones. Anyway, the pilot got a message from the ship to return ASAP because the weather was deteriorating rapidly. We quickly returned to the wide deck of the *San Martin*. Landing was smooth and we hardly felt our touchdown on the deck. Upon arrival, we were advised to remove our jumpsuits quickly, being careful not to walk on and tear the fabric on the foot part of the suit. If we damaged the feet of the suits they would leak if we landed in the freezing water.

Lamb and Waterhouse in helicopter jumpsuits.

Shortly, "The Horse" returned, circled the ship, and made a smooth landing as well. Lamb and Waterhouse had a successful trip and collected many species of lichens

126

from a nearby mountain peak, with an elevation of some 10,000 feet, where no person had ever set foot before.

After a rough night sleeping with lots of pitching and rolling, we arrived back at Palmer Station. Lamb and Zimmermann conferred with Mr. Austin, the NSF representative, and discussed our collecting plans for the next two days. Then Lamb and Zimmermann processed the abundant lichen collection and packed the specimens for return to Melchior.

Waterhouse and myself, in the meantime, were working on a dredge for collecting deep-water algae. It was designed and built in the *Edisto*'s machine shop. The dredge looked great and it had survived some testing earlier that afternoon. Several species of algae were collected at this location.

Later that day Zimmermann was invited to a small room on the ship where he met Jim the radioman who was talking on his amateur radio station. He had made contact with a party in Laredo, Texas. Jim asks the operator in Laredo if the telephone operators there are still on strike. "How does a Naval Officer in Antarctica know about that?" the Texas operator asks. Jim explains that the ship has a small, mimeographed daily newspaper. An agency in New York transmits news every night to ships all over the world. Jim just happened to be on duty when the message about the striking Laredo operators arrived via radio. They wanted higher pay because they had to be bilingual.

Port Arthur to Port Lockroy

The next stop was Port Lockroy, a British base where Mack Lamb was stationed for two years in 1944 and 1945. He was a member of the British "Operation Tabarin" a part of the British Antarctic Survey. The purpose of Operation Tabarin was to watch for German ships or submarines that could hide behind icebergs. Everyone was anxious to visit this historic, unoccupied base. Several officers from the ship, everyone in our group, and a navy photographer disembarked. Lamb ceremoniously knocked on the door of the unoccupied building. The space inside had been left since last utilized in 1961-62. Although Mack helped build this station when stationed here, the inside was totally unfamiliar to him. Many renovations and re-arrangement of rooms and furniture had been made. Even new walls had been built and old ones torn down or moved.

Zimmermann and I made a series of quick dives exploring the shallow, ice scoured shore to a depth of about 50 feet. The weather remained sunny, clear, and wind free, so we took some time to swim around and up on some small icebergs, acting like sleepy Weddell seals. Eventually we finished diving and returned to the ship.

At dinner Capt. Nickerson and Mr. Austin, from NSF, invited all the sailors and Dr. Bellisio at Melchior to visit the *Edisto* to see a movie, eat some ice cream and popcorn, and see how an American ship operates. The members of our team allowed each Argentine sailor to purchase $2.50 worth of merchandise. For the most part, the sailors chose an *Edisto* inscribed Zippo cigarette lighter and a carton of American cigarettes. After the shopping spree, the Argentine sailors along with the *Edisto* crew watched the movie. *Richard the Lion Hearted.* None of the Melchior crew understood a word of the dialogue, but they enjoyed the film anyway.

After the movie, I took Paulito, our cook, on a personal tour of the mess of the ship, where the American chef gave Paulito some milk, Ketchup, and other American condiments. Paulito was really impressed by both the 'toast making machine' and the nearby sit-down toilets for sailors (enlisted men) to use whenever they wanted. A latrine on the Argentine ship amounted to a room with a six-inch diameter hole in the floor. Another interesting thing was the free water. Any person could take a shower whenever he wished. Desalination was the secret, which the Argentine navy did not yet have on their icebreaker.

We were very late getting back to our base. Lamb lost track of his briefcase containing all his money, valuable papers, and records. The briefcase was finally recovered after several trips, about a mile away and back on the LCVP (landing craft for vehicles and personnel). When we finally returned to Melchior, the second generator gave its dying moans and was now completely out of commission.

We slept late and arose at 11AM and began cleaning up the diving gear, which we had unloaded at night in the dark. We were now seeing short (about 6 hours) night times. All tanks had to be filled and all neoprene suits dried, sorted, and organized. *Edisto* provided us with the use of a 26-foot motor whaleboat to help with our work, for the next week or two. This boat allowed us to increase the range of our diving and collecting excursions.

On February 14, it was cloudy and I got up at noon after staying up until 3AM finishing Stevenson's *Master of Bellantre*. From 1:30 on I had to read by flashlight because the generators were shut down early. Lamb made an early dive to look for *Ascoseira*, a brown kelp like algae, for Delépine. That night I developed films with Zimmermann. I was learning how to develop black and white film. We were able to develop some duplicate photographs, which we distributed among the support crew.

Almirante Brown

At Almirante Brown, there was no beach area to collect algae. The shoreline consisted of large volcanic boulders and volcanic substrate, and Lamb and I made an effort to snatch some specimens using a knife tied to a stick. The water was exceptionally clear, but I could not pry off and collect any complete specimens of algae, only fragments that were too small to have any value to our collections. The entire alga is required to be considered a complete specimen.

Port Lockroy and Almirante Brown Stations at the end of our season in Antarctica.

Chapter 8: Diversions

"I know not all that may be coming, but be it what it will, I'll go to it laughing."
-Herman Melville

THE DAY-TO-DAY LIFE organizing and carrying-out specific diving and algae collections in Antarctica was indeed exciting and worthwhile. However, following the same ritual over and over again started to take its toll. Everyone on the Harvard team had plenty of challenges maintaining interest in the work at hand. For me, I continued an interest in algae that started when I first went to work with Dr. Lamb. The daily routine became quite familiar four or five weeks into the expedition. Occasional diversions were in order, as well as a few pranks.

Special Celebrations

We always looked for an excuse to celebrate, and a person's birthday was cause for special celebration. We

132

discovered that December 1st was the birthday of René Delépine, age 31. Lt. Cueli had a vermouth and hors d'oeuvres table set up before lunch in honor of the occasion, and everybody at the base gathered together in the mess hall to drink to his good health and wish him many happy returns.

Christmas, December 25, 1964 was declared a day of rest. Katy had sent an artificial tree and materials that we could use to make ornaments.

Lamb's diary gave an engaging account of what we did for New Years Day.

January 1, 1965

A special New Year feast was prepared this evening by Hipolito Ojeda and his assistant Maximo Agudo. Extra chairs and table were taken into the living room to accommodate the throng, and at about 11 PM, all eighteen of us took our places at the table for dinner. Plates of colorful hors-d'oeuvres were passed around, with jugs of punch containing chopped apples a beverage known as "clerico" (white sangriå). The main course consisted of roast chicken, which came as a surprise, since few of us knew that we had

fresh, frozen chickens among our supplies in the refrigerator hut. But, further surprises were yet to come, as Ojeda then brought in and laid before our astonished eyes freshly baked iced cakes made in the form of

New Year's Day feast with the entire crew.

rabbit's heads and of a locomotive drawing a wagon filled with candy.

Lt. Cueli made a short speech congratulating all present on their achievements during this Antarctic campaign and wishing them a Happy New Year. Waterhouse expressed the sentiments of our scientific group and our appreciation of the links of lasting friendships which have developed during our stay at the Argentine base."

Guitars were then brought out and a musical fiesta was started which continued into the year 1965.

Orange Monkey Scurvy Prevention

Although the research, diving, and collecting algae, preparing specimens, and identifying unique species were interesting, there were still moments of boredom now and then. We cannot do scientific work every moment of the day so we had to fill in the gaps by creating and participating in diversions. Sometimes we could read, write in our journals, or study marine charts for future collecting excursions.

After an early dinner one night, a dessert of oranges was served. They are important because they keep well in storage and biologically they help prevent scurvy. Professor Zimmermann took out his pocketknife and began to carve something on the skin of his orange. When his carvings were finished he gently peeled the skin off the orange and we could see a scrambled-up ball of orange peel. The entire fistful of peel remained intact.

Zimmermann popped open a segment at the top of the orange and looped it over the neck of a nearby wine bottle. Once attached to the bottle, the ball of orange peel unraveled and the shape of a long-tailed monkey emerged. This was fascinating to all who had been watching his diligent work. We were all laughing and decided to give such a creative activity a try.

Needless to say, this event put a cheerful end to an otherwise boring day. I am glad to say none of us ever got scurvy but I noticed lots of orange peels around the base that looked like an orange monkey junkyard.

Bay of Relief

An entire afternoon was devoted to crossing the Melchior Sound to collect lichens from the rocky sidewalls of a fjord. Lamb, Zimmermann, Waterhouse, and I were in our small 12 foot, open research vessel. We made some snorkel dives to collect a few more specimens of marine algae.

Later we climbed around the rock walls nearby. Our curiosity was satisfied when we discovered small patches of moss plants wedged into small crevices between rocks. We also collected some more samples of the only vascular plant species growing on the Antarctic Continent, *Deschampsia Antarctica*, a cold tolerant grass species. The moss and grass specimens were bagged, labeled, and placed in a larger bag for return to our lab.

How do moss spores reach Antarctica and how do those spores find a place to settle and germinate? The spores are blown to Antarctica from continental South America in high speed, high elevation wind. The spores hit rocky outcroppings where V-shaped spaces had been created. Particles of soil and minerals from rocks weathering from above accumulate in those small microhabitats. The excrement from coastal bird droppings landing on the rocks above dribble into these nutrient rich soils and plants grow.

We all boarded the boat, started the outboard motor, and headed the 4 or 5 miles back to Melchior. The journey was slow and several of us had to urinate. We pulled off to our right, headed into a small, unnamed bay. We got out of the boat and swam around urinating in our neoprene wet suits, as was the custom; this also assured us a shower in the evening. The bay became known as the Bay of Relief and the cartographer was to have it placed on the Argentine Naval chart.

Zimmermann's Homemade Kites

One day that we were not completely buried with work, Martin decided to make a kite to see if the wind would take it afar. He gathered together a pile of trash from a wastebasket. The apparent trash consisted of several sheets of flimsy gift-wrapping tissue, a few metal coat hangers, some small pieces of wire and twine, and some Duco cement. Martin cut, twisted and bent the coat hangers into the shape of an elongated oval with a tapered rim at the base. From top to bottom measured about 2 feet. He then added a few more coat hanger ribs to secure his aircraft.

Next, after securing the ring like base, which was about 15 inches in diameter, he added another set of wire structures, which divided the base into four small chambers. He then attached a small metal jar cover to the center of the cross with the inner end of the jar cover facing upwards. He proceeded to apply the tissue paper around the wire frame, attaching the paper with Duco cement. It was almost finished and looked a lot like a Chinese celebration lantern.

We took the lantern outside where Martin placed some small, 1-inch pieces of cotton rope and a small amount of diesel fuel into the attached jar cover. Then Martin lit the fuel and waited for the heat to fill the kite that he made. Ever so slowly the hot air balloon/kite rose toward the blue sky. We watched the balloon drift in the breeze for about 20 minutes, until it disappeared from sight. It was a good thing we were so isolated since this would not be a smart idea near other buildings. Warning—do not try this at home or in any populated areas.

I was amazed that I had such a pleasant afternoon and glad that I learned firsthand how to make a hot air kite. Who knows if I will ever need this skill, but it certainly was fun.

Sometimes we just goofed off and took pictures of our antics.

Chapter 9
Leaving Antarctica

"The thing that is most beautiful about Antarctica for me is the light. It is like no other light on Earth, because the air is so free of impurities."
- Jon Krakauer

IN LATE FEBRUARY, we started planning and arranging our exit from Melchior Base. We had been here working since early November, some four months so far. We still had a number of places we would have liked to visit to expand our collection of algae, mosses, and lichens. But reality required precise scheduling, sorting, and distributing specimens. Most were dried on herbarium sheets but some specimens were preserved in Formalin for future microscopic study either in Paris with Delépine or in Cambridge with Lamb.

While we were starting to pack some of our gear one morning, the rompehielo *San Martin* made a surprise visit. They told us that the Argentine Minister of Defense would be arriving on the ship *Bahia Aquirre*. I expected some mail from Katy as well, and Zimmermann had given me some more penguin photos to send to Katy.

March 15 was the projected date to finish our research at Melchior and return to Buenos Aires. Once there, my plan was to visit Dr. Ernesto Foldats at the Universidade Central in Caracas, Venezuela. I sold most of my remaining film (Kodachrome, 22 rolls) to Waterhouse for 33 American dollars. I was low on cash and I expected to stay in Buenos Aries for about a week. It was a pleasure to visit old friends on the *San Martin* especially Lieutenants Olivera and Doncel.

The *Bahia Aguirre* finally arrived with the Minister of Defense, Dr. Leopoldo Suárez. Many photographs were taken. I made two radio interviews on tape, one for Radio Mundo Argentina and one for Radio Belgrano. We eventually got our mail, which included 21 letters from Katy. I decided to open the most recently cancelled postmark first. This was letter #21 (Katy numbered each envelope) and this confirmed that we were still engaged and she had not run away with my untrustworthy collections of close friends. I also received a letter from Frank Sanger, owner of the New England Divers in Beverly,

Massachusetts who had been so helpful equipping us for diving in the cold Antarctic waters.

During the *San Martin* icebreaker visit on February 16 and 17, we got the word from Captain Bustamante that the ship would return to Melchior to evacuate our base in early March. This meant that we had to prepare all of our equipment including scientific, diving SCUBA tanks, regulators, compressor as well as specimens of algae both dried and those preserved in Formalin, a standard biological preservative. This announcement would require us to prioritize our remaining collecting dives and allow enough time to clean and prepare specimens and equipment for shipping back to the United States. We also had to consider how the cargo was to be addressed. Some material had to go to the Natural History Museum in Paris, France and some to Harvard University in Cambridge Massachusetts USA. We also had to keep some clothing with us for travel back to the USA.

The formation of pack ice usually began in the first week of March. The days began getting slightly cooler and the daylight was beginning to give away to night time once again.

A few days later, Charlie Moorehead, piloting the *Edisto* helicopter, dropped in for a visit around 10 am. After having some freshly baked rolls and rotten coffee, they were ready to leave, but the Bell helicopter wouldn't start because the battery was low. Several tries to start the Bell were unsuccessful and a battery was requested by radio from the ship. Within the hour, a battery arrived and the helicopter returned safely to the *Edisto*.

Plans and deadlines were formed for our departure and everyone seemed upbeat at dinner. The cook roasted a pig. This meat was a feast. Imagine getting tired of eating prime Argentine beef, which we had eaten twice a day since we opened the base. After a fantastic dinner, we played cards with the sailors and also played guitars and sang folk songs. It was interesting to listen to Paulito play and sing 'musica folklorica' in his native Guaranee. He was still learning Spanish and had even improved his cooking.

Dr Bellisio tried to steal the large photograph of a penguin that Zimmermann had made for the dining room. When caught, he quietly sneaked away. But, he got revenge by blocking the doorways to our rooms by piling heavy packing crates so we could not enter. We all moved the crates, but got even with Bellisio by moving all the crates in front of his door once he had gone to sleep—we could hear him snoring loudly.

We were acting like kids in summer camp, ready to go home.

One of the last nights, I showed the Argentine sailors how to make a fire by rubbing some sticks together using a bow and dowel on a piece of pine. This is something every American Boy Scout learns as a Tenderfoot. All the sailors were very interested and waited anxiously to give it a try.

That night, I took my last shower in Antarctica. I washed a few clothes and piled most of my remaining clothes in a corner awaiting the final clothes-burning bonfire.

After lunch the next day, at 4:30 PM, Cueli, Lamb, Zimmermann and I went by boat to the other side of Gamma Island, the large island to which Melchior is attached, to collect soil samples, moss plants, and any lichens we could find. We easily found soil samples and some moss specimens. I really wondered how moss plants got to Antarctica in the first place. It finally became obvious that soil had accumulated in the small rock crevices on angular slopes.

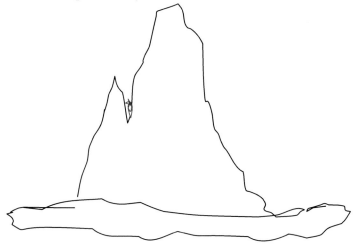

Crevices in the rocky slope formed small, sheltered pockets of soil, providing a microhabitat for moss plants. The soil, which was like fine sand, came from rock erosion from the area above the crevice. Dust from microscopic particles in the air, and perhaps some protein-based nutrients from bird guano, had all dissolved into the small crevices. There even appeared to be some small shell fragments from mollusks dropped there by birds. Moss spores carried aloft by air currents were deposited in these crevices and germinated.

I collected several moss specimens for a friend at the British Museum. As time progressed, I discovered a grass plant *Colobenthos Antarctica*, another example of a vascular plant growing on the Antarctic Peninsula.

Upon returning to the base we saw the icebreaker *San Martin* dropping anchor just outside the caleta. Our cargo was already prepared and loaded onto the ship. Of course, Zimmermann and I put a number of whale vertebrae in with the cargo for souvenirs for our friends back home. We quickly gathered our possessions and the ship's launch brought us out to the ship. We had dinner with the officers and were assigned to our cabins.

As we were leaving Melchior for the last time, we took note of the base which had now been cleared of snow, and took some pictures. We were sad to leave the beautiful snow-capped mountains and the vast expanse of barely studied marine coastline. Also, as we walked away from the base we saw the final pile of ashes where we had burned most of our clothes and shoes because they were so dirty and smelly.

Before we had left the base, via the casa de emergencia we all contributed some gifts (required by Antarctic tradition) for the next party to occupy this base. I left some toothpaste and a fifth of Drambuie liqueur. Each member of our party made substantial contributions to the treasure trash barrel to be discovered in the future.

Once on board the ship, we all showered and got cleaned up for lunch. At around noontime we arrived at Brown Station, named after Almirante Brown an Argentine naval hero. This station was occupied by an Argentine naval crew including a medical doctor who was quickly launched out to our ship. After a short conference with our ship's doctor, the ship moved to a calm cove nearby and another appendectomy was performed. The surgery was successful and the sailor walked up two flights of ship stairs the next morning to call his parents on the ship's radio. We knew him pretty well because he was the ship's diver. He was also the only diver I had ever seen using an obsolete Monsom lung for his diving work. This was like a life vest with a canister of oxygen attached. It looked like something from a bad science fiction movie.

Heading to Cape Horn

The trip to Cape Horn seemed like it would take an eternity. I scoured the ship for anything in English that was readable. All that I could find was a very old outdated copy of Esquire magazine and a History of Peruvian Indians. We all went to the ship's doctor to be weighed and measured. My circumference was considerably smaller than at the beginning of the expedition. My weight was down by 40 pounds and I felt great.

The helicopter pilot, Carlos Marioni, told me about the superstition regarding penguins. They were completely taboo on Argentine navy ships. When penguins were brought on board, the ship immediately encounters bad luck, storms, or even death. This was borne out during a campaign when a sailor on the ship *Bahia Aquirre* was killed by accident... he had brought a penguin on board. Another sailor got his hand caught in a large winch and supposedly lost a few fingers.

We stopped briefly at Deception where we had our last good meal before we departed for the Drake Passage and Cape Horn. We left Deception at 4:30 AM. It was not an easy trip. My journal entries from that trip describe clearly the challenges we faced as we tried to make our way back to Ushuaia, then Buenos Aries.

March 4, 1965

No entry! We were in the Drake Passage high seas, heavy rolling; our cabin was flooded because the porthole was left open by mistake.

March 5, 1965

Rough seas, we are now on course 358° almost due North. We are now headed to Puerto Belgrano to unload the two Helos. Today I shaved my moustache and got a haircut from Lt. Doncel, my first in five months.

March 9, 1965

Impossible to sleep last night, the ship was rolling from side to side, up to 60°. No surprise, icebreakers are designed much like a bathtub. The bow is blunt and the bottom of the ship is smooth and rounded. There is almost no keel. When the ship is breaking ice it is driven up on the ice and the weight of the ship crushes the ice into pieces. The process is incredibly noisy and no one can sleep, nor can you even carry on a conversation.

Back to the ship rolling last night... Many people were seasick, much of the furniture and dishes were broken. People ate fresh apples for nourishment since it was impossible to cook. We will enter Rio de la Plata at 2 AM tomorrow and after 22 hours navigating the river, arrive at Buenos Aires. There are lots of shallow parts in the channel. Everyone is very excited, but tired and aching from the three nights and days of rolling.

March 10, 1965

Yesterday we reached the mouth of the river at about 4 PM, but could not pick up a pilot because the sea was too rough for him to board the icebreaker. The Rio de la Plata (River of Silver) is over 100 miles from the open ocean.

Finally we arrived at Buenos Aires, and tied up at the dock, Darsena A, where everyone was lined up to see their families and friends. Our group stayed on board for a few more hours because none of us had families or friends waiting.

March 11, 1965

We had a great dinner that night, followed by a walk around the dock. Delépine was the first to leave; he was headed to Paris to see his family. His daughter was sick so he was anxious to go home. We had already dropped Dr. Lamb at Ushuaia where he planned to collect more lichen specimens. Richard Waterhouse was making some arrangements to have a sailboat built and he planned to sail it back to New England. Zimmermann and I got airline tickets. He was headed the next day to the Harvard Forest Lab in Petersham, Massachusetts. I was headed to Universidad Central in Caracas, Venezuela to visit an old friend Dr. Ernesto Foldats, a professor with whom Lamb and I had a student exchange between students from the algae course at Harvard and the students of an algae class at the Universidad Central in Caracas. Dr. Foldats showed me the beautiful coastline of Venezuela and many surrounding sights. His wife helped me shop for wedding rings.

March 16, 1965

I was headed back to Boston at last! My plane trip took a day and a night, stopping at numerous places, including Uruguay, Rio de Janeiro, Recife, Bélem, Miami, and New York. New York was interesting—the man sitting beside me was Red Auerbach, coach of the Boston Celtics (a favorite team of mine) and all his players. They had just lost an important game with New York and the players were all sleeping. I finally arrived at Boston's Logan Airport late on March 17th, Saint Patrick's Day, and borrowed some change from Auerbach to call Katy.

Saint Patrick's Day was a great day to be back home. I was completely happy to be on solid soil once again. The unstable feeling I had for two days after I left the San Martin was now gone. But, I found Boston traffic scary and it seemed that everyone drove too fast. I had not moved faster than a slow crawl for about five months. I also did not like the noisy world of Boston and Cambridge. The sounds of police cars, fire trucks, and ambulances were non-stop.

It was fun to be back with Katy and several friends, but strange that they seemed to talk more than necessary. All my friends wanted to know what the Antarctic experience was really like.

It was great to sleep in my own bed again. It felt much more comfortable than being in a sleeping bag on a bunk with an air mattress. The smells around me were familiar, but different. Most of all I was happy to get night time back.

Chapter 10
Epilogue

"The sea, once it casts its spell, holds one in the net of wonder forever."
Jacques Cousteau

AT THE END OF THE TRIP, most of us parted ways, each in pursuit of his own goals. The team leader, Dr. Ivan Mackenzie Lamb, Director of the Farlow Herbarium and Library at Harvard University, stayed in this position until he retired in 1977. Lamb continued working on the Antarctic algae collections as well as the lichens he had collected. He passed away in 1990 at the age of 79 from Lou Gherig's disease.

Dr. Martin H. Zimmermann, Bullard Professor of Forestry at Harvard, also continued his botanical work at Harvard. He passed away of cancer at the young age of 57.

Richard E. Waterhouse left our research group at the end of the expedition in Buenos Aires. He was planning to build a sailboat in Buenos Aires and sail it back to the United States, though I never found out if he accomplished this. I heard he taught in a New England prep school. He did not keep in touch with the team.

I came home and got married in June of 1965; Lamb and Zimmermann were guests at our wedding. They encouraged me to finish college; I could not have had better mentors.

I received my undergraduate degree at Salem State University, taught biology at a preparatory high school and became a graduate student of Marine Algae at the University of New Hampshire and received my doctorate in 1973. I taught at Plymouth State University, for 34 years, where I was able to conduct research on marine algae in Florida and the Virgin Islands living in underwater habitats. I studied and conducted research at Woods Hole Oceanographic Institution (Environmental Systems Laboratory), the University of the Azores, University of the Philippines, and Universidad Central in Venezuela. I maintain an interest in the physiology and ecology of algae throughout the world. In retirement, I read algal literature on a daily basis and try to convince my grandchildren to become botanists.

Around the late 1990's I received a letter from Dr. Betty Landrum of the Smithsonian Oceanographic Sorting Center asking about the Antarctic specimens that had been sent to Delépine in Paris. After we returned home, the many specimens of algae we had collected were transported back to Harvard and the National Museum in Paris. The collections were separated into two categories and some went to each laboratory for future study. Lamb had since retired from Harvard and no one at the Farlow Herbarium knew anything about the specimens.

Dr. Landrum asked me to make a visit with Delépine and obtain, if possible, the remaining Antarctic algae and return the collection to the Smithsonian. During the same trip, Landrum requested that I make a similar visit to Dr. Ludwig Salvini Plowen of the National Museum of Vienna, Austria. This assignment was to collect some overdue specimens of deep ocean invertebrates. That trip was quick and easy, though I felt a bit like a bill collector.

A few days later I was at Delépine's lab in Paris. He knew in advance that I was coming to collect the Antarctic algae specimens. This was an important collection and the only collection of this magnitude that existed in the world. Delépine sadly parted with the specimens. I had to buy two large, heavy-duty suitcases to transport the specimens back to Washington DC. Eventually, the entire collection was reunited and is now stored at the Farlow

Herbarium of Harvard University. This location serves as a final location of the Smithsonian specimens.

What Could Be Today

The following are a few observations to help collect more data using modern technological tools. For example, I wish I had collected samples of coastal phytoplankton, but none of our team even gave a thought to bringing a plankton net and preservative materials, like formalin and lugols solution. Also better charts, GPS, a good depth sounder and reliable marine radios and satellite cell phones for each person and small cell phones to call or check in at our base or take quick pictures, would have been helpful. Digital cameras, a GoPro, and drones would have been invaluable.

For diving we were actually in good shape for our early diving explorations. There have been advances in regulators, although the DA Aquamaster regulators would be perfectly safe today. I prefer Scubapro single hose regulators with a back-up second stage, and a dive computer with pressure gauge.

Many cold-water divers have learned that ¼ inch neoprene wet suits are warm near the surface of cold icy water. But as you dive deeper the neoprene cells compress with each additional atmosphere of pressure. By the time you reach a 33 foot depth, the cells of the quarter inch thick neoprene diving suits are equal to the thickness and insulating properties of 1/8-inch thick suits. The pressure squeezes the cells. That provides about as much warmth as a wet "T" shirt and thus limits bottom time to just minutes. In addition, your hands and feet become very cold quickly. Dry suits are a necessity.

Another important safety feature is dive boats. I recommend inflatable boats such as the 15-foot Zodiac to support three divers, one dive tender and all the diving equipment.

Modern day, automatic inflating life vests should be worn during any activities on the water. Divers in neoprene could be exempt, as they would float.

It was truly amazing that we had no serious physical or safety problems throughout the expedition. All of us were both sad to leave Melchior and yet anxious to return to our families and friends in Cambridge.

I look back on what we had for equipment and how we lived and feel that our expedition was closer to Shackleton's than any of today's scientific exploits.

A tremendous amount of scientific knowledge has been discovered in Antarctica during the post 50 years. Each year more new projects evolve.

One such project is the discovery of cryoconites, which develop when small soil particles accumulate on the upper surface of snow protrusions. The particles, like the oil particles at Melchior absorb radiation from the sun, melt some of the snow and develop microhabitats for algae and bacteria.

Potential Impacts of Global Climate Change (Global Warming)

A researcher named James/Joseph Farman collected records of ozone levels over Antarctica since 1956. Ozone, located 8-10 miles in the atmosphere absorbs ultra violet radiation (RES) from sunlight. By 1982 the values that Farman had collected showed a dramatic decrease in the ozone layer over Antarctica, particularly around the Antarctic Peninsula. The ozone hole or thin layer had been defined. This is going to cause major concern to shallow water algae and plankton. The Antarctic food chain is completely dependent on the filtration of RES—radiant energy from the sun—by the layer of ozone that protects the shallow water organisms from U.V. light. The cycle of 24 hours of daylight for 6 months of the year may have a serious impact on the growth, reproduction, and overall primary productivity of shallow water organisms. Future research could use our research as a baseline to evaluate damage to *in situ* algal populations by increases in UV light.

Interestingly, some researchers such as James McClintock 2012, Klaus Luning 1990, and Richard Moe 1977 have continued the research on these unique species of Antarctic algae.

Despite all of the discoveries that have taken place in Antarctica, people are really attracted to the natural beauty of this unique continent. I realize that modern technology can overcome most of the problems of tourist trash, organic pollution, and contamination of the atmosphere from tourist ships. I do however; have deep concerns over the possibility of building tourist hotels in such pristine, untouched areas as the Antarctic Peninsula.

Expedition Importance

The Harvard Botanical Survey of the West Antarctic Peninsula was important for many reasons. The first reason was this was one of the earliest opportunities to utilize a new research tool, SCUBA, to actually visualize what was going on with algae *in situ*. We knew how to use this tool from experiences in Rockport, Massachusetts on a short-term basis. We were able to study the seasonal succession of algae as well as their growth and reproduction. We had learned how to work together, as a team, to accomplish underwater algal studies. During this early period using SCUBA, we developed techniques for collecting, preserving, and maintaining laboratory cultures of reproductive structures. We developed several underwater cameras and experimented with 3/16" and ¼" neoprene diving suits and how to modify them for cold water diving. We also developed an underwater recording system to evaluate changes in underwater algal communities.

When we finally went to West Antarctica after a short course at the US Navy Diving School, we were as well prepared as possible. We worked with the best equipment available at the time.

I did not realize some of the constraints that we were to encounter. The island was exceptionally small with no opportunity to go for a walk. We could actually travel further by snorkeling than we could by walking. The wooden boats provided were such a small size we could not travel to sites we could easily see from land. When we were using the boats in open water it was easy to get caught by wind or by strong ocean currents. Technology has now overcome many of the limitations that we were exposed to in the 1964-1965 austral season.

We all had our individual moments where we could have a quiet day if needed. We all learned from each other how to cope with each other. Lamb had learned his coping skills long before during his two years stay at Port Lockroy in 1944 -1945, so he was well seasoned for our expedition. Zimmermann, Waterhouse, and I were new to the concept of isolation and tedious day after day searching for seaweeds.

I was duly impressed with our leader, Dr. Lamb. I often saw him study algae in the lab most of the day then make a dive, collect more algae, record his observation in his detailed journal, read some literature and finally go to bed late at night. Next day was a repeat of the day before. For Lamb, he seemed as if he had died and went to algae heaven. It was the same with Zimmermann, who could work on algae, dive, journal, do more algae work, and finally to sleep, only to repeat this schedule day after day—except Sundays, which were designated as a day of rest, unless of course the weather was calm and sunny, perfect for a collecting excursion.

Waterhouse and I were left to our own devices to combat boredom. However, I began an algae collection of my own, stemming from bits and pieces left over from the bulk collections available after each dive. I learned how to identify algae and how algal nomenclature was applied to each species. I learned how to observe growth stages and reproductive stages and how to evaluate algal habitats and the impacts of animal grazing. I read the basic literature about algae (there was nothing else left to read) and I participated in discussions about algal distributions and different population structure in cold, temperate and tropical conditions.

In some respects, I had a five month course you might call "Biology of the Algae," a course that I eventually taught as a professor at Plymouth State University.

Waterhouse did a lot of planning and organizing the logistics to send all our materials back to Cambridge. He organized the cargo and prepared inventories of all the equipment. He arranged materials to be left at Melchior for future explorers. He also began planning to have a sailboat built and sail back to the Boston area.

For me, the Antarctic Expedition provided a focus for the rest of my life. I did "buckle down," married Katy, finished college, taught in a prep school for two years, and got accepted to the University of New Hampshire under Dr. Arthur Mathieson, who continued to point me in the right direction as far as algae are concerned.

I've conducted research at Woods Hole Oceanographic Institution with Dr. John Ryther, studied mollusk and algae interaction in the Azores with Dr. Ruth Turner, and worked on algae from underwater habitats Tektite and UNH Edalhab, as well as acted as a consultant to the U.N. Development Program in Qatar and the Smithsonian Institution in the Philippines.

The Antarctic Expedition was important to all who participated. But, I learned a tremendous lesson about how different countries can do research together even when they are at different ends of the political spectrum. The organization of our expedition crossed political lines that tolerated each other even to the point of putting lives at risk to save each other.

At Melchior, every person that lived there would not hesitate for a moment to rescue someone in trouble. I can safely say that each of us felt that way as we reached the end of our work together. For each of us, this experience shaped our lives.

Appendix A
Antarctic Treaty (1959)

THIS IS A STREAMLINED VERSION of the main features of the Antarctic Treaty. The treaty was developed in 1959 and signed by the governments of Argentina, Chile, the French Republic, Japan, New Zealand, Norway, The Union of South Africa, the then USSR, the United Kingdom and the United States of America. "Recognizing that it is in the interest of all mankind that Antarctica shall continue forever to be used exclusively for peaceful purposes and shall not become the scene or object of international discord."

The treaty is built around international cooperation in scientific investigation in Antarctica. There are fourteen articles that constitute the body of the present treaty. They are presented here briefly.

Article 1. Antarctica shall be used for peaceful purposes only.

Article 2. Freedom of scientific investigation and cooperation in Antarctica.

Article 3. Plans for scientific programs and the exchange of scientific personnel shall be shared. Scientific results shall be exchanged and made freely available. Working relations with Specialized Agencies of the United Nations shall be established.

Article 4. Nothing in the present treaty shall be interpreted as influencing territorial sovereignty in Antarctica.

Article 5. Nuclear explosions and disposal of radioactive waste material is prohibited.

Article 6. The provisions of this treaty apply to the area south of 60 degrees.

Article 7. In order to promote the objectives and provisions of this treaty each member may designate observers to carry out any inspections concerning this treaty. This shall include all areas of Antarctica such as stations, installations, ships, aircraft, etc.

Article 8. Jurisdiction of personal and staff shall be subject only to the Contracting Party of which they are nationals.

Article 9. This article relates to periodic meetings of the Contracting (national) Parties to review measures regarding:

a) use of Antarctica for peaceful purposes
b) scientific research in Antarctica
c) international cooperation in Antarctica
d) rights of inspection in Antarctica

Article 10. Contracting Parties to make appropriate efforts, that no one engages in any activity in Antarctica contrary to the principals or purposes of the present treaty.

Article 11. If any dispute arises between two or more Contracting Parties the Parties shall consult among themselves to settle by peaceful means. If such a dispute is not settled it shall be referred to the International Court of Justice.

Article 12. The present treaty may be modified or amended at any time by unanimous agreement of the Contracting Parties (brief version).

Article 13. The present treaty shall be subject to ratification by the signatory States. It shall be open for accession by any State which is a member of the United Nations.

Article 14. The present treaty, done in the English, French, Russian, and Spanish languages, each version being equally authentic, shall be deposited in the archives of the Government of the United States of America.

Done at Washington DC this first day of December 1959.

As of 2016, 48 nations have agreed to the Antarctic Treaty. More recent years have included cooperative studies on climate change, pollution, Antarctic natural disasters, and enhancing tourist guidelines. Consultative meetings occur each year and the treaty is improved as needed.

Appendix B
Maps

This is the bottom part of the globe, showing Antarctica and the southern tip of Chile. The red dot shows approximately where Gamma Island lies.

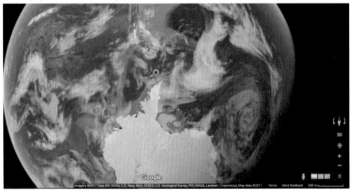

This aerial image shows the southern tip of South America and the northern part of the Antarctic Peninsula.

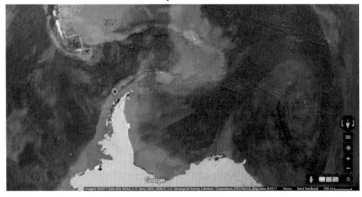

This aerial photo shows Gamma Island, and the location of Destacamento Melchior. The lines in the photo are due to satellite imagery that is pieced together.

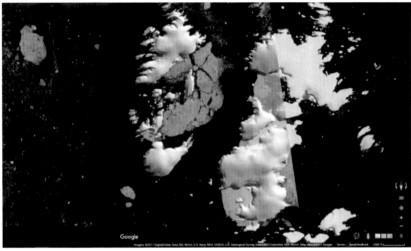

The red line in the photo below shows the route of the San Martin from Buenos Aires, stopping at Ushuia, and then travelling onward to the South Shetlands, Deception, and Melchior.

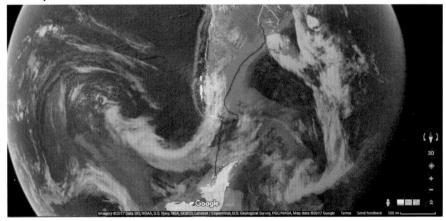

Appendix C
Glossary of Spanish Terms

aeroporto	airport
balenas	whales
bombillo	a metal straw for drinking matté
caldera	volcanic crater or a water heater boiler
caleta	small harbor
cama	bed
camarote	cabin
carne asado	roast beef
clerieb	Argentine beverage
destacamento	base, place where troops are stationed
empanada	deep fried meat pie
fango	mud
fumaroles	smoke, sulfuric gas from edges of an active volcano
guardia de agua	sailor or scientist who maintains daily water supply from shoveling snow to melt
Guaranee	a language or dialect spoken in mountain regions of northern Argentina and Paraguay
higado	liver
luna de miel	honeymoon
marinero	sailor
matté	tea made from yerba matté leaves
novia	girlfriend
rasca su bolas	scratch your balls
refugio	refuge, shelter, hotel
rhomephielo	ice breaker
salida de baño	bathrobe
Rio do la Plata	silver river
Tempano	iceberg
Tierra del Fuego	land of fire

Appendix D
Bibliography

Alexander, Caroline. The Endurance: Shackleton's Legendary Antarctic Experience. New York, NY: Alfred A. Knopf, 1998.

Armstrong, Jennifer. Shipwreck at the Bottom of the World. New York, NY: Crown Publishers, Inc., 1998.

Barnes, David, and Lloyd Peck. "Vulnerability of Antarctic shelf biodiversity to predicted regional warming." Climate Research 37 (October 2008): 149-63.

Cousteau, Jacques-Yves. The Whale: Mighty Monarch of the Sea. New York, NY: Doubleday and Company, Inc., 1972.

Day, David. Antarctica: A Biography. Oxford: Oxford University Press, 2013.

Dawes, Clinton J. Marine Botany. New York, NY: Wiley, 1981.

Delépine, R., I. Mackenzie Lamb, and M. H. Zimmermann. "Preliminary Report on the Marine Vegetation of the Antarctic Peninsula." Proceedings of the Fifth

International Seaweed Symposium, August 1966, 107-16.

Dewey, Jennifer Oswings. Four Months at the Bottom of the World. New York, NY: Scholastic Publishers, Inc., 2001.

Fralick, R. A., and A. C. Mathieson. "Physiological ecology of four Polysiphonia species (Rhodophyta, Ceramiales)." Marine Biology 29, no. 1 (1975): 29-36.

Fralick, R. A. Unpublished Personal Journal. Personal Collection of Richard Fralick. 1965 – 1966.

Lamb, I. M. Unpublished Personal Journal. Personal Collection of Richard Fralick. 1965 – 1966.

Lamb, I. M., and M. H. Zimmermann. "Marine Vegetation of Cape Ann, Essex County, Massachusetts." Rhodora 66 (1964): 217-54.

Lamb, I. Mackenzie, and Martin H. Zimmermann. "Benthic Marine Algae of the Antarctic Peninsula." American Geophysical Union Antarctic Research Series, 23 (4) (1977): 130-229.

Lüning, Klaus. Seaweeds: Their Environment, Biogeography and Ecophysiology. Wiley and Sons, Inc., 1990.

McClintock, James. Lost Antarctica: Adventures in a Disappearing Land. New York, NY: Palgrave Macmillan, 2012.

Moe, R. L., and P. C. Silva. "Antarctic Marine Flora: Uniquely Devoid of Kelps." Science 196, no. 4295 (1977): 1296-308.

Neushul, M. 1965a: Scuba diving studies of the vertical distribution of benthic marine algae. Acta Univ. Gothoburg. 3, 161-76.

Orr, James, et. al. "Anthropogenic ocean acidification over the twenty first century and its impacts on calcifying organisms." Nature 437 (2005): 681-86.

Salomon, David. Penguin-pedia: Photographs and Facts from One Man's Search for the Penguins of the World. Dallas, TX: Brown Books Pub Group, 2011.

Trewby, Mary, ed. Antarctica: An Encyclopedia from Abbott Ice Shelf to Zooplankton. Toronto: Firefly Books, Inc., 2002.

US Navy. "Bulletin." VI, no. 7 (June & July 1965): 28.

Vaughn, David. "Climate change on the Antarctic Peninsula." Climate Change 60 (2003): 243-74.

Watson, George E. "Birds of the Antarctic and Sub-Antarctic George E. Watson Birds of the Antarctic and sub-Antarctic." American Geophysical Union 18, no. 113 (1975).

Zimmermann, M. H. Unpublished Personal Journal. Personal Collection of Richard Fralick. 1965 – 1966.

Acknowledgements

First, I would like to thank Ariele Sieling, my editor and critic, who was outstanding in keeping me going with this story.

Thanks to the many people who were most important in my working life: I. M. Lamb, Martin Zimmermann, Ruth Turner, Arthur C. Mathieson, and John Ryther. Also, thanks to Sr. Mary Padraic, Saint Clement High School, Paul Nagy, New England Divers, Sylvia Earle, Manuel Marquez-Sterling, who were early influences in my life and choices I made.

Thanks to Bill Lane for digitizing my old slides for the pictures in this book. Special thanks to the readers and listeners who made valuable contributions to the book: Chris Bagnell, Steve Fink, Finn Fralick, Oliver Fralick, Bergan Garrity, Casey Garrity, Steve Habif, Ann Marie Jones, Bob Pollard, Chris Schultz, Linda Tatarczuch, and the Writing Group of the Active Retirement Association.

About the Author

Richard A. Fralick graduated from Salem State University with a BA degree in History and Biology. His M.S. and Ph.D. were in Plant Biology from the University of New Hampshire. At UNH he studied under Arthur Mathieson in the field of marine and freshwater algae. His major research was in the area of physiological ecology of *Codium fragile*, an invasive species of green algae and also the physiological ecology of four species of the red alga, *Polysiphonia*, distributed from open Atlantic Ocean to Great Bay Estuary in New Hampshire.

As an undergraduate he served as the chief technical diver on the Harvard University Expedition to Antarctica. At UNH he was a saturation diver on the Tektite Expedition in the Virgin Islands and also on the Edalhab FLARE expedition in the Florida Gulf stream. During his 34 years as a professor of Biology, at Plymouth State University, he worked summers on algal aquaculture under John Ryther at Woods Hole Oceanographic Institute (WHOI) and as a researcher for the U.S. Agency for International Development (USAID) on over harvesting mitigation for agar bearing red seaweeds in the Azores, Portugal. He has also done consulting for the

United Nations Development Program (UNDP) on algal problems in the Persian Gulf, and algal consulting in the Philippines. Dick was also a Smithsonian Consultant on Antarctic Algae and medical, anti-cancer drugs extracted from algae. More recently he worked, jointly with the Farlow Herbarium at Harvard University, on an algal database for N.E. Marine Algae funded by NH Sea Grant and on a Curriculum Project funded by the National Fisheries Institute. He has published numerous articles and technical reports often with students, concerning results of his work with marine algae. Dick is on the board of the Blue Ocean Society for Marine Conservation.

He currently lives in Durham, New Hampshire with his wife. They spend time sailing, skiing, traveling, going to UNH hockey games, grandparenting, and still, collecting algae.

KEEP ANTARCTICA COOL.

Made in the USA
Middletown, DE
31 March 2017